INTERNATIONAL CENTRE FOR MECHANICAL SCIENCES

COURSES AND LECTURES - No. 17

ANGELO MARZOLLO
UNIVERSITY OF TRIESTE

CONTROLLABILITY AND OPTIMIZATION

LECTURES HELD AT THE DEPARTMENT
FOR AUTOMATION AND INFORMATION
SEPTEMBER - OCTOBER 1969

UDINE 1969

SPRINGER-VERLAG WIEN GMBH

Originally published by Springer-Verlag Wien New York in 1972

ISBN 978-3-211-81123-8 ISBN 978-3-7091-2959-3 (eBook)

DOI 10.1007/978-3-7091-2959-3

Preface

In Chapter I of these lecture notes functional analysis methods will be used to derive in detail the conditions of controllability of continuous linear time-varying systems, for the case in which the controls are bounded in norm by a given constant, the time interval is not only finite but also fixed, and the controllability is intended as the possibility of transferring the state vector from one given point $\underline{x}_0$ to an other given point $\underline{x}_1$ of state space. These conditions are both more general and someway of more practical interest than the usual controllability conditions in which initial and final points, as well as the time interval, are fixed, and the controls are not norm-limited.

From the controllability conditions, as exposed here, it is easy to derive conditions of optimality (both in time and in norm), as it will be done in Chapter II, where for the special case of controls belonging to an Hilbert space the completely explicit form of optimal controls will be given. An exercize will clarify some delicate points of the theory, particularly for the case in which the system is not controllable in the usual sense of the term, in which case time optimal controls may be not norm minimal.

The main result of Chapter III is the

formal solution of the problem of "controllability in the presence of noise", which is a problem analogous to the one of Chapter I, but with noise present (and the point $\underline{x}_1$ substituted with a given region A of the state space).

Chapter I and II, as well as the first part of Chapter III are essentially an elaboration of Professor H.A. Antosiewiez's material, already appeared in [1].

Only the formal solution of the problem of controllability in the presence of noise, which constitutes the second part of Chapter III, is original; it should be pointed out, however, that the method and the main tool used (separation theorem for convex compact sets) were inspired by Professor Antosiewiez's idea for the solution of the controllability problem in the absence of noise.

I wish to acknoledge Dr. S. De Julio for the very useful exchange of ideas on the material exposed in Chapter I and II; I wish also to thank CISM for the invitation to held this course and its students for their attention and helpful suggestions.

Angelo Marzollo

Udine, October 1969

chapter 1

CONTROLLABILITY

Let us consider a linear control system, described by a linear differential equation of the type

$$\dot{\underline{x}} = A(t)\underline{x} + B(t)\underline{u} . \qquad I(1)$$

In I(1), and throughout these notes, $\underline{x}(t)$ is an n-component vector, $A(t)$ is an $n \times n$ matrix and $B(t)$ is an $n \times m$ matrix such that on the bounded interval $K = [t_0, t_1]$ the elements $a_{ij}(t)$ of $A(t)$ are Lebesgue measurable and the elements $b_{ij}(t)$ of $B(t)$ are Lebesgue integrable with their q^{th} power ($1 \leq q < \infty$), and $\underline{u}(t)$ is an m-component vector whose components are L-integrable with their p^{th} power ($1 \leq p < \infty$) on the same interval K. Moreover, the class $\mathcal{U}$ of "admissible control functions" will consist for the moment of functions $\underline{u}$ such that $\underline{u}(t) \in Q$ a.e. in K, where Q is an appropriate convex compact set in the m dimensional space to which $\underline{u}(t)$ belongs. Further requirements on $\mathcal{U}$ will be made later. As it is well known, given $K = [t_0, t_1]$ and any $\underline{x}_0 \in R^n$, for each $\underline{u} \in \mathcal{U}$ the equation I(1) has in K an unique (and absolutely continuous) solution $\underline{v}(t)$, with $\underline{v}(t_0) = \underline{x}_0$, which is given by

$$\underline{v}(t) = V(t)\underline{x}_0 + \int_{t_0}^{t} V(t)\, V^{-1}(s)\, \underline{u}(s)\, ds \qquad I(2)$$

where $V(t)$ is the solution of the matrix equation

$$\frac{dV(t)}{dt} = A(t)\,V(t)$$

with initial value $V(t_0) = I$

To stress the dependance of $\underline{v}$ on $\underline{u}$, we shall sometimes write $\underline{v}_{\underline{u}}(t)$ instead of $\underline{v}(t)$. The initial time will always be understood to be t_0 even when not explicitely mentioned.

Let us define as "region reachable at t_1, from $\underline{x}_0$" the set $R_{\underline{x}_0}(K)$ of points $\underline{x} \in R^n$ such that $\underline{v}_{\underline{u}}$ given by I(2) in correspondence to an admissible control $\underline{u}$ is $\underline{x}$ at t_1. $R_{\underline{x}_0}(K)$ is therefore defined by

$$R_{\underline{x}_0}(K) = \left\{ \underline{v}_{\underline{u}}(t_1)\,,\ \underline{u} \in \mathcal{U} \right\}$$

If we consider a non empty set $A \subset R^n$, the distance Δ between the sets $R_{\underline{x}_0}(K)$ and A is defined as

$$\Delta\left[R_{x_0}(K)\,,\,A\right] = \inf_{\underline{x} \in R_{\underline{x}_0}(K),\,\underline{y} \in A} d(\underline{x},\underline{y})$$

where $d(x,y)$ is the usual Euclidean distance between the points x and y of R. It is clear by the definition of $R_{x_0}(K)$ that the distance Δ is equal to the infimum

$$\inf_{\underline{u} \in \mathcal{U},\,\underline{y} \in A} d\left[\underline{v}_{\underline{u}}(t_1), \underline{y}\right] = \inf_{\underline{u} \in \mathcal{U}} \delta\left[\underline{v}_{\underline{u}}(t_1), A\right] \qquad \text{I(3)}$$

which means that Δ is the infimum of the distance δ from A of all admissible trajectories at t_1, starting from $\underline{x}_0$. The first problem we are going to face is the following: does there exist in the stated hypothesis an admissible control function $\underline{u}_0$ such that the corresponding infimum of I (3) is also a minimum? In other words, the question is whether there exists an admissible control $\underline{u}_0$ such that the corresponding value of $\underline{v}_{\underline{u}_0}(t)$ at t_1 has a minimal distance from A among all $\underline{v}_{\underline{u}}(t_1)$, $\underline{u} \in \mathcal{U}$. In Theorem 3 of this chapter it will be shown that the answer to this question is in the affirmative sense, and such an admissible control function exists. In the preliminary Theorems I 1 and I 2 the convexity and compactness of the reachable set $R_{\underline{x}_0}(k)$ will be proved, under the already stated hypothesis on $\mathcal{U}$. These important properties of $R_{\underline{x}_0}(k)$ are basic for the solution of an other problem, this one of great practical importance: for sets A and classes $\mathcal{U}$ of special kind, is it possible to find conditions in order that the distance I(3) is zero or, in other words, in order that A is reachable at t_1 by an admissible trajectory?

With respect to this last problem, we shall consider $\underline{u}$ to belong to specific linear normed spaces and consist of controls $\underline{u}$ bounded in norm by a given constant ϱ, and we shall consider A to be either the closed ball $A_\varepsilon(\underline{x}_1)$ of radius ε around a given point $\underline{x}_1$, or the convex compact set $A^n_\varepsilon(\underline{x}_1)$ which we shall define in chapter III for treating the case of presence of noise as input, or to be simply a given point $\underline{x}_1$. Although this last case may be obviously considered as a particular case of the other ones, it will be treated first, and Theorem I,4 which will appear at the end of this chapter will give necessary and sufficent conditions on the system I(1), and on $\underline{x}_0, \underline{x}_1$, $K = [t_0, t_1]$, ϱ for x_1 to be reachable by an admissible trajectory.

A generalization of Theorem I,4 will be given in Theorem III,1, where the case $A = A_\varepsilon(\underline{x}_1)$ will be considered, and the result will be reached with a slightly different proof, which could indeed be used also to prove Theorem I,4 .

The following Theorem I,1 and I,2 may be considered as preliminary and will be used also in Chapter 3.

Theorem I, 1

In the already stated hypothesis for system I(1) and admissible class $\mathcal{U}$ of controls, the set $R_{\underline{x}_0}(K)$ is convex.

Proof

From I(2), it is clear that the set $R_{\underline{x}_0}(K)$ is given by

$$R_{\underline{x}_0}(K) = \left\{ V(t_1)\underline{x}_0 + \int_{t_0}^{t_1} V(t_1)\, V^{-1}(s)\, B(s)\, \underline{u}(s)\, ds \,,\ \underline{u} \in \mathcal{U} \right\} \tag{I(4)}$$

If we define the map Λ_K from $\mathcal{U}$ to R^n as follows:

$$\Lambda_K(\underline{u}) = \int_{t_0}^{t_1} V(t_1)V^{-1}(s)\,B(s)\,\underline{u}(s)\,ds \qquad \forall \underline{u} \in \mathcal{U} \qquad \text{I(5)}$$

we need to prove only that the set $\Lambda_K(\mathcal{U})$ is convex, since $R_{\underline{x}_0}(K) = V(t_1)\,\underline{x}_0 + \Lambda_K(\mathcal{U})$ and $V(t_1)\underline{x}_0$ is a fixed point.

To show that $\Lambda_K(\mathcal{U})$ is convex, suppose $\underline{u}_1$, and $\underline{u}_2$ to be two arbitrary points of $\mathcal{U}$, so that $\Lambda_K(\underline{u}_1)$ and $\Lambda_K(\underline{u}_2)$ are two arbitrary points $\Lambda_K(\mathcal{U})$. Then, for $0 \leq \mu \leq 1$ we have

$$\mu\Lambda_K(\underline{u}_1) + (1-\mu)\,\Lambda_K(\underline{u}_2) = \Lambda_K\left[\mu\,\underline{u}_1 + (1-\mu)\,\underline{u}_2\right]$$

since the map Λ_K is obviously linear.

Furthermore, since the functions $\underline{u}_1$, and $\underline{u}_2$ belong to $\mathcal{U}$, $\underline{u}_1(t)$ and $\underline{u}_2(t)$ belong to Q for $t \in K$ a.e., and by the convexity of Q also $\mu u(t) + (1-\mu)\;u_2(t)$ belongs to Q for $t \in K$ a.e., and therefore $\mu\,\underline{u}_1 + (1-\mu)\,\underline{u}_2$ belongs to $\mathcal{U}$.

Hence $\Lambda_K\left[\mu\,\underline{u}_1 + (1-\mu)\,\underline{u}_2\right]$ belongs to $\Lambda_K(\mathcal{U})$ and $\Lambda_K(\mathcal{U})$ is convex, since $\Lambda_K(\underline{u}_1)$ and $\Lambda_K(\underline{u}_2)$ were arbitrary, and the proof is completed.

Let us now take the usual Euclidean norm for the control vector $\underline{u}$:

$$\| \underline{u}(t) \| = \left\{ \sum_{1}^{m}{}_i \, | u_i(t) |^2 \right\}^{1/2}$$

and since the admissible controls $\underline{u}$ of $\mathcal{U}$ are such that the p-power $\| \underline{u}\,(t) \|^p$ is $\mathcal{L}$-integrable in K, for any integer p greater then 1, we may define the norm $\| \underline{u} \|_p$;

$$\| \underline{u} \|_p = \left[\int_{t_0}^{t_1} \| \underline{u}(s) \|^p \, ds \right]^{1/p} \qquad 1 < p < \infty \qquad \text{I(6)}$$

The space of functions $\underline{u}$ equipped with the norm I (6) is defined as the space $\mathcal{L}_{2,p}$ of m-vector functions.

It may be shown that the space $\mathcal{L}_{2,p}$ is a reflexive Banach space, since $\mathcal{L}_p (1 < p < \infty)$ is. (Actually, this is true for any $\mathcal{L}_{r,p}$ space, with $1 \leq r \leq \infty$, $1 < p < \infty$, cfr. for example [2] ch. 4). In next Lemma I,1 and Theorem I,2, the set $\mathcal{U}$ will be considered as belonging to $\mathcal{L}_{2,p}$.

We shall now prove Lemma I,1, which will be useful for the proof of the following important theorem I,2.

<u>L e m m a I,1</u>

Λ_K defined in I(5) is a continuous map from $\mathcal{L}_{2,p}$ to R^n.

Proof

Taking the Euclidean norm of both sides of I(5) and using obvious inequalities we have

$$\|\Lambda_K(\underline{u})\| = \left\| \int_{t_2}^{t_1} V(t_1) V^{-1}(s)\, B(s)\, \underline{u}(s)\, ds \right\| \le \int_{t_0}^{t_1} \| V(t_1) V^{-1}(s) B(s) \underline{u}(s) \| \, ds \le$$

$$\int_{t_0}^{t_1} \| V(t_1) V^{-1}(s) B(s) \| \, \| \underline{u}(s) \| \, ds , \qquad \text{I(7)}$$

where $\| V(t_1) V^{-1}(t) B(t) \|$ is the norm of the $n \times m$ matrix $V(t_1) V^{-1}(t) B(t)$ defined as the $\sup\limits_{\underline{y} \in R^m} \dfrac{\| V(t) V^{-1}(t) B(t) \underline{y} \|}{\| \underline{y} \|}$

Using Hölder's inequality we have

$$\int_{t_0}^{t_1} \| V(t_1) V^{-1}(s) B(s) \| \, \| \underline{u}(s) \| ds \le \left(\int_{t_0}^{t_1} \| V(t_1) V^{-1}(s) B(s) \|^q ds \right)^{1/q} \left(\int_{t_0}^{t_1} \| \underline{u}(s) \|^p ds \right)^{1/p}$$

where $\frac{1}{p} + \frac{1}{q} = 1 \quad , \; 1 < p \le \infty$. I(8)

Inequalities I(7) and I (8) give

$$\| \Lambda_K(\underline{u}) \| \le K \| \underline{u} \|_p$$

from which it is clear that Λ_K is a continuous map (in the metric or strong topology, defined by the norm $\| \underline{u} \|_p$) .

Theorem I,2

The set $R_{\underline{x}_0}(K)$ is compact.

Proof

It is clearly sufficient to show that $\Lambda_K(\mathcal{U})$ is compact. Let us first recall that given a linear space X, and therefore the linear space X^* of all linear functionals $\underline{f}$ defined on X, and a finite subset A of X^*, we can define the weak or X^* topology on X in the following way. If p is any point of X, the X^* topology is the one having as base all sets N of the form

$$N(p,A,\varepsilon) = \left\{ q : |\underline{f}(p) - \underline{f}(q)| < \varepsilon \ , \ \varepsilon > 0 \ , \ \underline{f} \in A \right\}$$

From the definition of the weak topology follows in a natural way the definition of weakly open, weakly closed, weakly compact sets, as well as of weak continuity.

Recall that (cfr. [3] p.422), since Λ_K is continuous, it is also weakly continuous, hence it maps weakly compact sets into weakly compact sets, and that Λ_K is a map from $\mathcal{U}$ into the n-dimensional space R^n, where weak and strong compactness coincide. We need therefore to show only that the set $\Lambda_K(\mathcal{U})$ is weakly compact.

In order to do this, let us first prove that it is closed in the strong topology, induced by the norm I(6). Let us suppose that $\underline{u}_1 \ldots\ldots \underline{u}_n \ldots$ is a sequence of functions belonging to $\mathcal{U}$ and converging to $\underline{u}$, i.e. let

$$\lim_{n\to\infty} \|\underline{u}_n - \underline{u}\|_p = 0$$

or

$$\lim_{n\to\infty} \int_{t_0}^{t_1} \|\underline{u}_n(s) - \underline{u}(s)\|^p \, ds = 0$$

Since $\| \underline{u}_n(t) - \underline{u}(t) \|$ is integrable, and non negative, we have

$$\lim_{n \to \infty} \int_{t_0}^{t_1} \| \underline{u}_n(s) - \underline{u}(s) \| ds = 0$$

Applying the well-known theorem of Lebesgue, for arbitrary $\varepsilon > 0$, as n tends to infinity the measure μ of the set S_ε of all t such that $| \underline{u}_n(t) - \underline{u}(t) | > \varepsilon$ tends to zero, or $\underline{u}_n$ tends to $\underline{u}$ in measure. Threfore there exists a subsequence $\underline{u}_{n_i}$, which converges almost everywhere to $\underline{u}$:

$$\lim_{n_i \to \infty} \underline{u}_{n_i}(t) = \underline{u}(t) \qquad \text{a.e. in } K$$

Since $\underline{u}_{n_i}(t)$ belongs to Q a.e. for $t \in K$, and Q is compact, hence closed, also $\underline{u}(t)$ belongs to Q almost everywhere in K, and we reach the conclusion that $\underline{u}$ belongs to $\mathcal{U}$, and $\mathcal{U}$ is closed.

Let us now notice that the space $\mathcal{L}_{2,p}$ $(1 < p < \infty)$ is obviously locally convex, i.e. the base for its topology consists of convex sets. As it was made clear also during the proof of Theorem 1, the class $\mathcal{U}$ is a convex set. Since, as we proved just now, the set $\mathcal{U}$ is closed in the (strong) topology induced by the norm I(6), if we consider $\mathcal{U}$ as a subset of the space $X = \mathcal{L}_{2,p}$, we can apply the following theorem, which we report from [3], p. 422 "A convex subset of a locally convex linear topological space is weakly closed if and only if it is (strongly) closed" and we may conclude that $\mathcal{U}$ is weakly closed.

It is clear that $\mathcal{U}$ is bounded, since Q is and since, as we mentioned, the space $\mathcal{L}_{2,p}(1<p<\infty)$ is a reflexive Banach space, we may apply the following theorem, which we report from [3] , p.425 "A bounded weakly closed set in a reflexive Banach space is weakly compact" and conclude that $\mathcal{U}$ is weakly compact, or compact in the X^* topology. The proof of theorem 2 is so completed: $\Lambda_K(\mathcal{U})$ is a compact set in R^n, and therefore also $R_{\underline{x}_0}(K)$ is.

Using theorem 2 it is now possible to get the answer to the first problem we posed at the beginning of this chapter. We put this answer in the form of the following theorem I,3

<u>Theorem I,3</u>

Given a non empty set $A \subset R^n$, there exists at least one $\underline{u}_0 \in \mathcal{U}$ such that

$$\delta\left[\underline{v}_{\underline{u}_0}(t_1), A\right] = \inf_{\underline{u}\in\mathcal{U}} \delta\left[\underline{v}_{\underline{u}}(t_1), A\right]. \qquad I(9)$$

<u>Proof</u>

The set $\underline{v}_{\underline{u}}(\underline{t}_1)$, with $\underline{u}$ ranging over $\mathcal{U}$, is the reachable set $R_{\underline{x}_0}(K)$. So

$$\inf_{\underline{u}\in\mathcal{U}} \delta\left[\underline{v}_{\underline{u}}(t_1), A\right] = \inf_{\underline{x}\in R_{x_0}(K)} \delta\left[\underline{x}, A\right].$$

By definition of infimum, there exists a sequence $\{\underline{x}_n\}$ in $R_{\underline{x}_0}(K)$ such that

$$\inf_{\underline{x} \in R_{\underline{x}_0}(K)} \delta\left[\underline{x}, A\right] = \lim_{n \to \infty} \delta\left[\underline{x}_n, A\right]$$

By definition of this distance $\delta\left[\underline{x}_n, A\right]$, we have $\delta\left[\underline{x}_n, A\right] = \inf_{\underline{y} \in A} d\left[\underline{x}_n, \underline{y}\right]$, and there exists a sequence $\{\underline{y}_j\}$ in A such that $\inf_{\underline{y} \in A} d\left[\underline{x}_n, \underline{y}\right] = \lim_{j \to \infty} d\left[\underline{x}_n, \underline{y}_j\right]$, and threfore

$$\inf_{\underline{u} \in U} \delta\left[\underline{V}_{\underline{u}}(t_1), A\right] = \lim_{n,j \to \infty} d\left[\underline{x}_n, \underline{y}_j\right]$$

Since $R_{\underline{x}_0}(K)$ is compact by theorem 2, every sequence belonging to it contains a subsequence with limit in $R_{\underline{x}_0}(K)$. So the sequence $\{\underline{x}_n\}$ contains a subsequence $\{\underline{x}_{n_K}\}$ such that $\lim_{n_K \to \infty} \underline{x}_{n_K} = \underline{x}^0 \in R_{\underline{x}_0}(K)$. Let the corresponding subsequence in A be denoted by $\{\underline{y}_{j_K}\}$. Then, since the function distance d is continuous, we have

$$\lim_{n,j \to \infty} d\left[\underline{x}_n, \underline{y}_j\right] = \lim_{n_K, j_K \to \infty} d\left[\underline{x}_{n_K}, \underline{y}_{j_K}\right] = \lim_{j_K \to \infty} d\left[\underline{x}^0, \underline{y}_{j_K}\right], \text{ hence}$$

$$\inf_{\underline{u} \in U} \delta\left[\underline{v}_{\underline{u}}(t_1), A\right] = \lim_{j_K \to \infty} d\left[\underline{x}^0, \underline{y}_{j_K}\right] \text{ and}$$

there exists in $R_{\underline{x}_0}(K)$ a point $\underline{x}^0$ of minimal distance from A and theorem I,3 is proved.

Notice that, if A were compact, then $\{\underline{y}_{j_K}\}$ would contain a convergent subsequence $\{\underline{y}_{j_K}\}$ with limit $\underline{y}^0$ in A and therefore the distance between sets $R_{\underline{x}_0}(K)$ and A would in this case be the distance $d(\underline{x}^0, \underline{y}^0)$ between points $\underline{x}^0$ and $\underline{y}^0$.

This of course is the case when A reduces to a point $\underline{x}_1$.

As we already mentioned, for the case of A reducing to a point $\underline{x}_1$ and special classes $\mathcal{U}$ of admissible controls, we shall now give necessary and sufficient conditions for the "state" vector $\underline{x}(t)$ of I(1) to reach $\underline{x}_1$ at t_1 starting from $\underline{x}_0$ at t_0.

Let us first give the following

DEFINITION

The system I(1) is called "controllable" relative to points $\underline{x}_0$, $\underline{x}_1$, time interval $K = [t_0, t_1]$ and the class $\mathcal{U}$ of admissible controls, if there exists at least one $\underline{u} \in \mathcal{U}$ such that the corresponding solution $\underline{v}_{\underline{u}}(t)$ starting from $\underline{x}_0$ at time t_0 reaches point $\underline{x}_1$ at time t_1.

For shortage, such a system will sometimes be called "$(\underline{x}_0, \underline{x}_1, \mathcal{U}_p^\varrho, K)$ controllable"

From now on, we shall consider the case in which the class $\mathcal{U}$ of admissible controls is the class $\mathcal{U}_p^\varrho$ of controls $\underline{u}$ such that the Euclidean norm $\|\underline{u}(t)\|$ of $\underline{u}(t)$ is Lebesgue p-integrable, and

$$\|\underline{u}\|_p = \left[\int_{t_0}^{t_1} \|u(s)\|^p \, ds\right]^{1/p} \leq \varrho , \quad 1 < p \leq \infty , \quad \varrho > 0 \tag{I(10)}$$

with the usual definition

$$\|u\|_{\infty} = \operatorname{ess\,sup}_{t \in K} \|u(t)\| . \qquad I(10')$$

We give now the main result of this chapter, in the form of the following

Theorem I,4

The system I(1) is ($\underline{x}_0 , \underline{x}_1 , K , \mathcal{U}_p^{\varrho}$) controllable if and only if for every $\underline{x} \in R^n$ the following inequality holds

$$< \underline{x}, \underline{x}_1 - V(t_1)\,\underline{x}_0 > \; \leq \varrho \left(\int_{t_0}^{t_1} \| B^*(s)\, V^{-1*}(s)\, V^*(t_1)\, \underline{x} \|^q ds \right)^{1/q} \qquad I(11)$$

where $p > 1$, and $\frac{1}{p} + \frac{1}{q} = 1$,

Proof

Necessity: suppose that there exists a control $\underline{u} \in \mathcal{U}_p^{\varrho}$ such that

$$\underline{x}_1 = V(t_1)\,\underline{x}_0 + \int_{t_0}^{t_1} V(t_1)\, V^{-1}(s)\, B(s)\, \underline{u}(s)\, ds .$$

Rearranging, taking the scalar product with $\underline{x}$, where $\underline{x}$ is any vector of $\mathbf{R}^n$, and defining for convenience

$$W(t_1, s) = V(t_1)\, V^{-1}(s)\, B(s) \qquad I(12)$$

and

$$\underline{z}(t_1) = \underline{x}_1 - V(t_1)\,\underline{x}_0 \qquad I(13)$$

we have

$$<\underline{x}, \underline{z}(t_1)> = <\underline{x}, \int_{t_0}^{t_1} W(t_1,s)\,\underline{u}(s)\,ds> \quad \forall \underline{x} \in \mathbb{R}^n$$

and therefore

$$|<\underline{x}, \underline{z}(t_1)>| = \left| \int_{t_0}^{t_1} <\underline{x}, W(t_1,s)\,\underline{u}(s)> ds \right| . \quad \forall \underline{x} \in \mathbb{R}^n$$

Defining $W^*(t_1,s)$ to be the adjoint of $W(t_1,s)$ we have

$$|<\underline{x}, \underline{z}(t_1)>| = \left| \int_{t_0}^{t_1} <W^*(t_1,s)\,\underline{x}, \underline{u}(s)> ds \right| \quad \forall \underline{x} \in \mathbb{R}^n$$

Using Holder's inequality and recalling the definition of $\|u\|_p$ we have

$$|<\underline{x}, \underline{z}(t_1)>| \leq \int_{t_0}^{t_1} |<W^*(t_1,s)\,\underline{x}, \underline{u}(s)>|\,ds \leq \|\underline{u}\|_p \left[\int_{t_0}^{t_1} \|W^*(t_1,s)\,\underline{x}\|^q\, ds \right]^{1/q}$$

and recalling the position I(12) the necessity is proved.

Sufficency

Suppose that I(11) holds and $\underline{x}_1$ cannot be reached from $\underline{x}_0$, which means that $\underline{x}_1$, does not belong to the reachable set $R_{\underline{x}_0}(K)$.

Let us first notice that the properties of $\mathcal{U}$ of being convex and (strongly) closed, which we used in Theorem I,1,2 to show that $R_{x_0}(K)$ was convex and compact are obviously enjoyed by $\mathcal{U}_p^e (1<p<\infty)$ considered as a subset of $\mathcal{L}_{2,p}$. For the case $\mathcal{U}_\infty^e$ corresponding to the norm I(10'), it is still obvious that $\mathcal{U}_\infty^e$ is convex; to show that it is (strongly) closed in the reflexive Banach space $\mathcal{L}_{2,p}$ $(1<p<\infty)$, notice that $\underline{u} \in \mathcal{U}_{\infty}^{\varrho}$, or $\| u \|_{\infty} \leq \varrho$, implies that $\underline{u}(t)$ belongs a.e. to an appro priate compact set Q . We may therefore use again Theorem 2 and also in this case $R_{\underline{x}_0}(K)$ is compact. Using this property of $R_{\underline{x}_0}(K)$ and the fact that $\underline{x}_1$ is supposed not to belong to it, we shall now prove that there exists an hyperplane strictly separating $R_{\underline{x}_0}(K)$ from $\underline{x}_1$, or that there exists a vector $\underline{x}'' = R^n$ and a constant α such that

$$< \underline{x}'', \underline{x} > \ > \alpha > \ < \underline{x}'', \underline{x}_1 > \qquad \forall x \in R_{\underline{x}_0}(K) \qquad \text{I(14)}$$

We shall first prove the following Lemma, which will be useful also in Chapter 3, when we shall prove that a convex compact set may be strictly separated by any convex closed set disjoint from it.

<u>Lemma I,2</u>

If T is a non empty closed convex set of R^n not containing the origin, then there exists a vector $\underline{x}'' \in R^n$ and a positive constant γ such that

$$< \underline{x}'', \underline{y} > \ > \gamma \qquad \forall \underline{y} \in T \qquad \text{I(15)}$$

<u>P r o o f</u>

Define the closed ball B in R^n:

$$B = \underline{x} : \|\underline{x}\| \leq \ell$$

whose radius ℓ is such that the intersection $I = T \cap B$ is non-empty. Since I is compact, the continuous function $f(\underline{x}) = \|\underline{x}\|$ with domain I takes its minimal value in a point $\underline{x}'' \in I$, so that

$$\|\underline{x}\| \geq \|\underline{x}''\| \quad \forall \underline{x} \in I$$

$$\|\underline{x}\| \geq \|\underline{x}''\| \quad \forall \underline{x} \in T \qquad I(16)$$

To prove now that

$$<\underline{x}'', \underline{y}> \;\geq \|\underline{x}''\|^2 = \gamma \quad \forall \underline{y} \in T \qquad I(15')$$

suppose that there exists an $\underline{y}'' \in T$ such that $<\underline{x}'', \underline{y}''> \, < \|\underline{x}''\|^2$, so that there exists a positive constant ε with

$$<\underline{x}'', \underline{y}''> = \|\underline{x}''\|^2 - \varepsilon \qquad I(17)$$

Consider, for t in the interval $0 \leq t \leq 1$, the quantity

$$\|(1-t)\underline{x}'' + t\underline{y}''\|^2 - \|\underline{x}''\|$$

Recalling I(17), we have

$$\|(1-t)\underline{x}'' + t\underline{y}''\|^2 - \|\underline{x}''\|^2 = (1-t)^2\|\underline{x}''\|^2 + 2t(1-t)<\underline{x}'', \underline{y}''> + t^2\|\underline{y}''\|^2 - \|\underline{x}''\|^2 =$$

$$= t^2\|\underline{x}''\|^2 - 2t\|\underline{x}''\|^2 + 2t(1-t)\left[\|\underline{x}''\|^2 - \varepsilon\right] + t^2\|\underline{y}''\|^2 =$$

$$= -2t(1-t)\varepsilon + t^2\left(\|\underline{y}''\|^2 - \|\underline{x}''\|^2\right).$$

For t small enough, the following inequality clearly holds

$$t^2\left(\|\underline{y}''\|^2 - \|\underline{x}''\|^2\right) < 2\varepsilon t(1-t)$$

and therefore clearly

$$\| (1-t)\underline{x}'' + t\underline{y}''\|^2 - \|\underline{x}''\|^2 < -2t(1-t)\varepsilon + 2\varepsilon t(1-t) = 0$$

Moreover, since T is convex and both $\underline{x}''$ and $\underline{y}''$ belong to it, also $(1-t)\underline{x}'' + t\underline{y}''$ does and therefore, by I(16)

$$\| (1-t)x'' + ty''\|^2 - \| x''\|^2 \geq 0$$

and we have reached a contraddiction. So Lemma I(2) is proved and I(15), I(15') hold for some $x'' \in R^n$.

Defining now the set C ·

$$C = R_{\underline{x}_0}(K) - x_1 = \left\{ \underline{\lambda} : \underline{\lambda} = \underline{x} - \underline{x}_1 , \underline{x} \in R_{\underline{x}_0}(K) \right\}$$

since we supposed that $\underline{x}_1$ does not belong to $R_{\underline{x}_0}(K)$, the set C does not contain the origin and is convex and compact. Therefore by Lemma I,2, for some $\underline{x}'' \in R^n$ we have

$$< \underline{x}'', \underline{x} > - < \underline{x}'', \underline{x}_1 > \geq \gamma > 0 \qquad \forall \underline{x} \in R_{\underline{x}_0}(K)$$

and therefore

$$\inf_{\underline{x} \in R_{\underline{x}_0}(K)} < \underline{x}'', \underline{x} > \; > \; < \underline{x}'', \underline{x}_1 > \qquad I(18)$$

and there exists a constant α such that I(14) is satisfied.

Considering the vector $\underline{x}' = -\underline{x}''$, from I(14) we have

$$< \underline{x}', \underline{x} > \; < -\alpha < \; < \underline{x}', \underline{x}_1 > \qquad \forall \underline{x} \in R_{\underline{x}_0}(K)$$

from which, putting $\beta = -\alpha - < \underline{x}', V(t_1)\underline{x}_0 >$, we have

$$< \underline{x}', \underline{x} - V(t_1)\underline{x}_0 > \; < \beta < \; < \underline{x}', \underline{x}_1 - V(t_1)\underline{x}_0 > \qquad \forall \underline{x} \in R_{\underline{x}_0}(K)$$

and therefore, since $\Lambda_K(\mathcal{U}) = R_{\underline{x}_0}(K) - V(t_1)\underline{x}_0$,

$$< \underline{x}', \underline{x} > \; < \beta < \; < \underline{x}', \underline{x}_1 - V(t_1)\underline{x}_0 > \qquad \forall \underline{x} \in \Lambda_K(\mathcal{U}) \qquad \text{I(19)}$$

The inequalities I(19) are obviously equivalent to

$$< \underline{x}', \int_{t_0}^{t_1} W(t_1, s)\, \underline{u}(s)\, ds > \; < \beta \qquad \forall\, \underline{u} \in \mathcal{U}_p^e \qquad \text{I(20')}$$

$$< \underline{x}', \underline{x}_1 - V(t_1)\, \underline{x}_0 > \; < \beta \qquad \text{I(20'')}$$

Inequality I(20') holds also if we take the absolute value of its first member, since if $\underline{u}$ belongs to $\mathcal{U}_p^e$, also $-\underline{u}$ does. So we have

$$\left| < x', \int_{t_0}^{t_1} W(t_1, s)\, \underline{u}(s)\, ds > \right| = \left| \int_{t_0}^{t_1} < W^*(t_1, s)\underline{x}', \underline{u}(s) > ds \right| < \beta$$

Taking the supremum of both sides we have

$$\sup_{\underline{u} \in \mathcal{U}_p^\varrho} \left| \int_{t_0}^{t_1} < W^*(t_1,s)\, \underline{x}', \underline{u}(s) > ds \right| \leq \beta \qquad \text{I(21)}$$

Notice now that $\int_{t_0}^{t_1} < W^*(t_1,s)\, \underline{x}', \underline{u}(s) > ds$ is a linear continuous functional $\underline{f}(\underline{u})$, with $\underline{u}$ belonging to the normed space $\mathcal{L}_{2,p}$. Therefore, putting $\underline{r}(s) = \frac{1}{\varrho}\, \underline{u}(s)$, we have

$$\sup_{\|\underline{u}\|_p \leq \varrho} \left| \int_{t_0}^{t_1} < W^*(t_1,s)\, \underline{x}', \underline{u}(s) > ds \right| = \varrho \sup_{\|\underline{r}\|_p \leq 1} \left| \int_{t_0}^{t_1} < W^*(t_1,s)\, \underline{x}', \underline{r}(s) > ds \right| =$$

$$= \varrho \sup_{\|r\|_p \leq \varrho} \left| \underline{f}(\underline{r}) \right| = \varrho \|\underline{f}\|_q \qquad \text{I(22)}$$

where (cfr. [2] ch. 4 for properties of functionals on vector valued function spaces)

$$\|\underline{f}\|_q = \left(\int_{t_0}^{t_1} \| W^*(t_1,s)\, \underline{x}' \|^q \, ds \right)^{1/q} \qquad \text{I(23)}$$

is taken in the space $\mathcal{L}_{2,q}$ conjugate of $\mathcal{L}_{2,p}$.

Putting together I(21),I(22) and I(23) we have

$$\beta \geq \varrho \left(\int_{t_0}^{t_1} \| W^*(t_1,s)\, \underline{x}' \|^q \, ds \right)^{1/q} \qquad \text{I(24)}$$

which, together with I(20"), says that there exists a vector $\underline{x}' \in R^n$ such that

$$|<\underline{x}', \underline{x}_1 - V(t_1)\underline{x}_0>| > \varrho\left(\int_{t_0}^{t_1}\|W^*(t_1,s)\,\underline{x}'\|^q ds\right)^{1/q} = \varrho\left(\int_{t_0}^{t_1}\|B^*(s)V^{-1*}(s)V^*(t_1)\underline{x}'\|^q ds\right)^{1/q}$$

which contraddicts the hypothesis that I(11) holds for every $\underline{x} \in R^n$.

So we have reached the conclusion that if I(11) holds, then the system I(1) is controllable relative to points $\underline{x}_0$, $\underline{x}_1$, time interval K and class $\mathcal{U}_p^\varrho$ of admissible controls, and Theorem I,4 is proved . Let us make some comments on it.

As mentioned, in Chapter III we shall prove a generalization of this theorem in Theorem III,1 which will give a condition similar to I(II), valid for the case in which $\underline{x}_1$ is substituted with a closed ball $A_\varepsilon(\underline{x}_1)$ around it. For that generalization we shall use again Lemma I,2, which was essential here.

For the moment, let us notice that we considered the norm of the m-dimensional vector $\underline{u}(t)$ to be Euclidean, (and we took $\|\underline{u}(t)\| = \left\{\sum_1^m {}_i\, |u_i(t)|^2\right\}^{1/2}$) only as an example. We could actually equip the space to which $\mathcal{U}$ belongs with any other norm, for example

$$\|\underline{u}(t)\|_r = \left\{\sum_1^m {}_i\, |u_i(t)|^r\right\}^{1/r} \qquad 1 \leq r < \infty$$

and

$$\|u(t)\|_\infty = \max_{i=1\ldots m} |u_i(t)| \qquad \text{for } r = \infty$$

and all previous statements would still hold, of course with a appropriate changes of norms. For example, in Theorem I,4 the norm of $B^*(t)V^{-1*}(t)V^*(t_1)$ would be taken in the space conjugate to the one in which $\underline{u}(t)$ is considered, so that by $\| B^*(t)V^{-1*}(t)V^*(t_1)\underline{x} \|$ we would mean $\{ \sum_1^m {}_i | B^*(t)V^{-1*}(t)V^*(t_1)\underline{x} |_i^s \}^{\frac{1}{s}}$ with $\frac{1}{r} + \frac{1}{s} = 1$.

This is understood also in the next chapter.

A more significant remark concerns the meaning of theorem I,4. Its importance actually lies in two facts: the first one is that, as we shall see in the following chapter, we can derive from it the existance and the form of optimal controls, i.e. controls which transfer the "state" vector $\underline{x}(t)$ from $\underline{x}_0$ to $\underline{x}_1$ in a minimal time with controls bounded in norm by a certain constant, and which make the same transfer with controls of minimal norm. The value of this minimal norm will also be computed for the case in which the norm is the "energy".

The second fact is that the notion of "controllability" as used in theorem 4 is more meaningful from a physical point of view and also more general that the usual notion of "controllability" of linear systems.

The "controllability" of linear time varying systems is usually defined as the existance of some control (with an <u>arbitrarly large</u> norm) which transfer <u>any</u> point $\underline{x}_0 \in R^m$ at

t_0 to <u>any</u> point $\underline{x}_1 \in R^m$ at an arbitrarily large, finite time.

We may observe that

a) in practize, there are obvious physical limitations on the controls which it is possible to implement and use as inputs to a physical system, and therefore a limitation on the norm of the admissible controls is appropriate.

b) the contrallibility within a <u>given</u> time interval is phisically more interesting than the controllability within a time interval which is required only to be finite

c) it may happen that we are interested not so much in the possibility of transferring the state vector from any point to any other point of the state space, but only from a definite point $\underline{x}_0$ to an other definite point $\underline{x}_1$. In this case, the usual definition of controllability is too stringent to be of practical use.

d) Theorem 4 is actually a generalization of the well known necessary and sufficent condition for controllability in the usual sense. This condition is the following
"A linear time varying system is controllable at time t_0 if and only if there exists a time T such that

$$Y(T) = \int_{t_0}^{T} W(t_1, s)\, W^*(t_1, s)\, ds$$

is positive definite"

We shall now prove that this condition is a special case of theorem I,4 when $\underline{u}$ is considered to belong to the normed space $\mathcal{L}_{2,2}$:

$$\|\underline{u}\|_2 = \left\{ \int_{t_0}^{t_1} \sum_{1}^{m} {}_i \left[u_i(s)\right]^2 ds \right\}^{1/2}$$

In theorem 1, take the class of admissible controls to be

$$\mathcal{U}_2^\varrho = \left\{ \underline{u} : \|\underline{u}\|_2 \le \varrho \right\}$$

The inequality I(11) becomes

$$\left| < \underline{x}, \underline{x}_1 - V(t_1)\underline{x}_0 > \right| \le \varrho \left[\int_{t_0}^{t_1} \| W^*(s)\,\underline{x} \|^2 \, ds \right]^{1/2} \quad \forall \underline{x} \in R^n \qquad \text{I(25)}$$

Using again position I(13), the inequality I(25) becomes

$$| < x, z(t_1) > | \le \varrho \left[\int_{t_0}^{t_1} < W(t_1,s)\, W^*(t_1,s)\; \underline{x}, \underline{x} > ds \right]^{1/2} \; \forall \underline{x} \in R^n$$

Notincing that

$$< \underline{x}, \underline{z}(t_1) > \; = \sum_{1}^{n} {}_i x_i z_i(t_1) \sum_{1}^{m} {}_j x_j z_j(t_1) = \sum_{1}^{n} {}_{ij} z_i(t_1) z_j(t_1) x_i x_j$$

defining the matrix $Z(t_1)$ as the matrix whose (i,j) element is $z_i(t_1)\, z_j(t_1)$ and taking the square of both members of I(26) we get

$$\int_{t_0}^{t_1} < W(t_1,s)\, W^*(t_1,s)\; \underline{x}, \underline{x} > ds - \frac{1}{\varrho^2} < Z(t_1)\underline{x}, \underline{x} > \ge 0 \quad \forall x \in R^n \qquad \text{I(27)}$$

From Theorem I,4 we have obtained the following

Corollary I,1

The system I(1) is controllable relative to $\underline{x}_0, \underline{x}_1$, time interval K and admissible controls $\underline{u} \in \mathcal{U}_2^{\varrho}$ if and only if the matrix

$$C(t_1) = W(t_1,s)\, W^*(t_1,s)\, ds - \frac{1}{\varrho^2} Z(t_1) \qquad \text{I(28)}$$

is positive semidefinite.

If in Corollary I,1 of theorem I,4 we suppose $\underline{x}_0, \underline{x}_1$, arbitrary and ϱ , t_1 arbitrarily large, we have the condition of controllability we mentioned at the beginning of point d).

Indeed, suppose $Y(t)$ to be positive definite for some T . Then, for any $\underline{x}_0$ and $\underline{x}_1$, i.e. for any $\underline{z}(T)$, it simply suffices to inspect $C(T)$ to verify the existance of a constant ϱ' such that, for $\varrho \geq \varrho'$, $C(T)$ is positive semidefinite.

Conversely, if we suppose that $C(t_1)$ is positive semi definite for all $\underline{z}(t_1) \in R^n$, then $Y(t_1)$ is positive definite.

Indeed since,

$$< Y(t_1)\, \underline{x}, \underline{x} > = \int_{t_0}^{t_1} < W(t_1,s)\, W^*(t_1,s)\, \underline{x}, \underline{x} > ds =$$

$$\int_{t_0}^{t_1} < W^*(t_1,s)\, \underline{x}, W^*(t_1,s)\, \underline{x} > ds = \int_{t_0}^{t_1} \| W^*(t_1,s)\, \underline{x} \|^2 ds \geq 0 ,$$

if $Y(t_1)$ was not positive definite, it would be singular, which is impossible since then for any $\underline{x}$ such that $Y(t_1)\underline{x}=0$ and for $\underline{z}(t_1)$ such that $< \underline{z}(t_1), \underline{x} > \neq 0$, we would have

$$< Y(t_1)\underline{x}, \underline{x} > - \frac{1}{\rho^2} < \underline{x}, \underline{z}(t_1) >< \underline{x}, \underline{z}(t_1) > < 0$$

that is

$$< Y(t_1)\underline{x}, \underline{x} > - \frac{1}{\rho^2} < \underline{z}(t_1)\underline{x}, \underline{x} > < 0$$

and $C(t_1)$ would not be positive semidefinite.

So theorem I,4 and therefore corollary I,1 are actually a generalization of the usual condition of controllability for a linear time varying system.

chapter 2

OPTIMIZATION

In this chapter we shall obtain results concerning the existance of time optimal controls as well as of norm optimal controls, we shall deduce a special case of Pontryagin maximum principle for linear systems, and we shall also find the time optimal and norm minimal control functions.

These results will be obtained assuming the controllability relative to given points $\underline{x}_0, \underline{x}_1$, a class $\mathcal{U}_p^{\varrho}$ of admissible controls and a given interval K, and not the usual hypothesis of controllability relative to arbitrary points, class of admissible controls, and time interval.

The results of the previous chapter, namely theorem I,4, will be heavily exploited. We shall exclude from our considerations the trivial case in which, if K is the given time interval, there exists $t^* \in K$ such that $\underline{x}_1 = V(t^*)\underline{x}_0$, and therefore the function $\underline{u} = 0$ effects the desired transfer from $\underline{x}_0$ to $\underline{x}_1$.

Let us first prove a theorem concerning time optimal controls.

Theorem II, 1

If the system I(1) is controllable relative to points $\underline{x}_0$ and $\underline{x}_1$, time interval $K = [t_0, t_1]$ and the class $\mathcal{U}_p^e$ of admissible controls, then there exists a least interval $\bar{K} = [t_0, \bar{t}\,]$, $(t_0 < \bar{t} \leq t_1)$ in which it is controllable. In other words, if a system is $(\underline{x}_0, \underline{x}_1, \mathcal{U}_p^e, K)$ controllable, then there exists a time optimal control.

Proof

Let us consider the set $H = \{t, t > t_0\}$ over which the system is $(x_0, x_1, \mathcal{U}_p^e, H)$ controllable. This set is clearly non-empty. Let us define

$$\bar{t} = \inf H \qquad \text{II(1)}$$

We need to prove that the system I(1) is controllable relative to $\underline{x}_0$, $\underline{x}_1$, $\mathcal{U}_p^e$ and $\bar{K} = [t_0, \bar{t}]$, i.e. the interval $\bar{K}$ is the minimal interval over which the system I(1) is controllable, which is equivalent to say that $\bar{t}$ belongs to H. Indeed, suppose that $\bar{t}$ does not belong to H. Then by theorem I,4 there would exist a vector $\hat{\underline{x}}$ in R^n such that

$$|< \hat{\underline{x}}, \underline{x}_1 - \underline{V}(\bar{t})\underline{x}_0 >| > \varrho \left(\int_{t_0}^{\bar{t}} \| B^*(s) V^{-1*}(s) V^*(\bar{t}) \hat{\underline{x}} \|^q \, ds \right)^{1/q} \qquad \text{II(2)}$$

By the definition II(1) of $\bar{t}$, there exists a sequence $\{t_n\}$ with $\lim_{n\to\infty} t_n = \bar{t}$ and such that for each interval $K_n = [t_0, t_n]$ the system is controllable; and therefore

$$|<\hat{\underline{x}},\underline{x} - V(t_n)\underline{x}_0>| \leq \varrho \left(\int_{t_0}^{t_n}\|B^*(s)V^{-1*}(s)V^*(t_n)\hat{\underline{x}}\|^q\right)^{1/q} \qquad \text{II(3)}$$

Defining

$$|<\hat{\underline{x}},\underline{x}_1 - V(t_n)\underline{x}_0>| = \gamma(t_n)$$

$$\varrho\left[\int_{t_0}^{t_n}\|B^*(s)V^{-1*}(s)V^*(t_1)\hat{\underline{x}}\|^q ds\right]^{1/q} = \delta(t_n)$$

and

$$|<\hat{\underline{x}},\underline{x}_1 - V(\bar{t})\underline{x}_0>| = \gamma$$

$$\varrho\left[\int_{t_0}^{\bar{t}}\|B^*(s)V^{-1*}(s)V^*(\bar{t})\hat{\underline{x}}\|^q ds\right]^{1/q} = \delta \qquad \text{II(4)}$$

since $\gamma(t_n)$ and $\delta(t_n)$ are continuous functions, we have

$$\lim_{n\to\infty} \gamma(t_n) = \gamma(\bar{t}) = \gamma \qquad \text{II(5')}$$

$$\lim_{n\to\infty} \delta(t_n) = \delta(\bar{t}) = \delta \qquad \text{II(5'')}$$

Inequality II(3) is equivalent to $\gamma(t_n) \leq \delta(t_n)$, and therefore

$$\lim_{n\to\infty} \gamma(t_n) \leq \lim_{n\to\infty} \delta(t_n)$$

and by II(5'), II(5") we have $\gamma \leq \delta$. Recalling the position II(4) we see that this last inequality contraddicts II(2), and Theorem II(1) is proved.

<u>Theorem II,2</u>

If $\bar{K} = [t_0, \bar{t}]$ is the minimal time interval over which the system I(1) can be transferred from $\underline{x}_0$ to $\underline{x}_1$ with an admissible control belonging to $\mathcal{U}_p^\rho$, then there exists a vector $\underline{x}' \in R^n$ such that

$$|<\underline{x}', \underline{x}_1 - V(\bar{t})\underline{x}_0>| = \rho \left(\int_{t_0}^{\bar{t}} \| B^*(s) V^{-1*}(s) V^*(\bar{t}) \underline{x}' \|^q ds \right)^{1/q} \qquad \text{II(6)}$$

<u>Proof</u>

Since the system I(1) is ($\underline{x}_0$, $\underline{x}_1$, $\mathcal{U}_p^\rho$, $\bar{K}$) controllable, the inequality

$$|<\underline{x}, \underline{x}_1 - V(\bar{t})\underline{x}_0>| \leq \rho \left(\int_{t_0}^{\bar{t}} \| B^*(s) V^{-1*}(s) V^*(\bar{t}) \underline{x} \|^q ds \right)^{1/q} \qquad \text{II(7)}$$

holds for all $\underline{x} \in R^n$.

Let us consider a sequence $\{t_n\}$ converging to $\bar{t}$, with $t_n < \bar{t}$. For each t_n , there exists a vector $\underline{x}_n$ such that

$$|<\underline{x}_n, \underline{x}_1 - V(t_n)x_0>| > \rho \left(\int_{t_0}^{t_n} \| B^*(s) V^{-1*}(s) V^*(t_n) \underline{x}_n \|^q \right)^{1/q} \qquad \text{II(8)}$$

The inequality II(8) clearly holds if we substitute $\underline{x}_n$ with $\lambda_n \underline{x}_n' = \underline{x}_n$ where $\|\underline{x}_n'\| = 1$. In this way, all $\underline{x}_n'$ belong to the compact unit sphere of R^n, and therefore from the sequence $\{\underline{x}_n'\}$ it is possible to extract a convergent subsequence $\{\underline{x}'_{n_K}\}$ such that $\lim_{n_K \to \infty} \underline{x}'_{n_K} = \underline{x}'$, with $\|\underline{x}'\| = 1$. Since both terms of inequality II(8) (with $\underline{x}_n'$ instead of $\underline{x}_n$) depend continuously on t_n and on $\underline{x}_n'$, then

$$\lim_{\substack{\underline{x}'_{n_K} \to \underline{x}' \\ t_{n_K} \to \bar{t}}} |<\underline{x}_n, \underline{x}_1 - V(t_{n_K})\underline{x}_0>| = |<\underline{x}', \underline{x}_1 - V(\bar{t})\underline{x}_0>|$$

and

$$\lim_{\substack{\underline{x}'_{n_K} \to \underline{x}' \\ t_{n_K} \to \bar{t}}} \left(\int_{t_0}^{t_n} \|B^*(s)V^{-1*}(s)V^*(t_n)\underline{x}'_{n_K}\|^q ds\right)^{1/q} = \left(\int_{t_0}^{\bar{t}} \|B^*(s)V^{-1*}(s)V^*(\bar{t})\underline{x}'\|^q ds\right)^{1/q}$$

and therefore, recalling the inequality II(8), we have

$$|<\underline{x}', \underline{x}_1 - V(\bar{t})\underline{x}_0>| \geq \varrho\left(\int_{t_0}^{\bar{t}} \|B(s)V^{-1*}(s)V^*(\bar{t})\underline{x}'\|^q ds\right)^{1/q}. \qquad \text{II(9)}$$

Confronting II(7) with II(9), it turns out that for $\underline{x} = \underline{x}'$, the equality II(6) holds and the theorem is proved.

We may put the result of theorem II,2 in a negative form, namely in the form of the following

Corollary II,1

If inequality I(11) holds as a strict inequality for all $\underline{x} \in R^n$, than t_1, cannot be the optimal time of transfer from $\underline{x}_0$ to $\underline{x}_1$.

Let us now turn to the problem of the existance of the so called "minimal norm" control: given the system I(1), which is supposed to be ($\underline{x}_0$, $\underline{x}_1$, $\mathcal{U}_p^\varrho$, K) controllable, does a control $\hat{\underline{u}}$ exist such that $\|\hat{\underline{u}}\|_p = \min_{u \in \mathcal{U}_p^\varrho} \|\underline{u}\|_p$, and which transfers $\underline{x}_0$ to $\underline{x}_1$, within K ?

The answer to this question is in the affermative sense, and the control $\hat{\underline{u}}$ is called "minimal norm" control. We are going to prove this fact in the following

Theorem II,3

Let the system I(1) be ($\underline{x}_0$, $\underline{x}_1$, $\mathcal{U}_p^\varrho$, $K = [t_0, t_1]$) controllable and let $\bar{\mathcal{U}} \subset \mathcal{U}_p^\varrho$ be the set of functions $\underline{u}$ which effect the transfer from $\underline{x}_0$ to $\underline{x}_1$ at time t_1, i.e. such that

$$\int_{t_0}^{t_1} V(t_1) V^{-1}(s) B(s) \underline{u}(s) ds = \underline{x}_1 - V(t_1)\underline{x}_0 \quad \text{if } \underline{u} \in \bar{\mathcal{U}}$$

Then there exists a function $\hat{\underline{u}}$ such that

$$\|\hat{\underline{u}}\|_p = \inf_{\underline{u} \in \bar{\mathcal{U}}} \|\underline{u}\|_p = \hat{\varrho} \qquad \text{II(10)}$$

Proof

Let us define $\hat{\varrho}$ to be the $\inf_{\underline{u} \in \bar{\mathcal{U}}} \|\underline{u}\|_p$. There exists therefore a sequence $\{\varrho_n\}$, with $\varrho_n > \hat{\varrho}$ and such that $\lim_{n \to \infty} \varrho_n = \hat{\varrho}$. For each ϱ_n the system is ($\underline{x}_0$, $\underline{x}$, $\mathcal{U}_p^{\varrho_n}$, K) controllable, so that for every $\underline{x} \in R^n$

$$|<\underline{x}, \underline{x}_1 - V(t_1)\underline{x}_0>| \leq \varrho_n \left(\int_{t_0}^{t_1} \| B^*(s) V^{-1*}(s) V^*(t_1) \underline{x} \|^q ds \right)^{1/q}.$$

This inequality is conserved as n tends to infinity, so we have

$$|<\underline{x}, \underline{x}_1 - V(t_1)\underline{x}_0>| \leq \hat{\varrho} \left(\int_{t_0}^{t_1} \| B^*(s) V^{-1*}(s) V^*(t_1) \underline{x} \|^q ds \right)^{1/q} \quad \forall \underline{x} \in R^n \qquad \text{II(11)}$$

and by theorem I(4) there exists a control function $\hat{\underline{u}} \in \bar{\mathcal{U}}$, with $\|\hat{\underline{u}}\|_p \leq \hat{\varrho}$. Since $\hat{\underline{u}} \in \bar{\mathcal{U}}$, and $\inf_{\underline{u} \in \bar{\mathcal{U}}} \|\underline{u}\|_p = \hat{\varrho}$, $\|\hat{\underline{u}}\|_p$ must be equal to $\hat{\varrho}$, and the infimum in II(10) is assumed.

We want now to give an expression of $\hat{\varrho}$ as a function of the parameters of the system I(1), and of $\underline{x}_0, \underline{x}_1$, t_0 and t_1. This expression is given in the following

Theorem II,4

Using the definition of theorem II(3) for $\bar{\mathcal{U}}$, we have

$$\hat{\varrho} = \inf_{\underline{u} \in \bar{U}} \|\underline{u}\|_p = \sup_{\left(\int_{t_0}^{t_1} \|W^*(t_1,s)\underline{x}\|^q ds\right)^{\frac{1}{q}} = 1} < \underline{x}, \underline{x}_1 - V(t_1)\underline{x}_0 > \qquad \text{II(12)}$$

P r o o f

Since II(11) is satisfied, we have

$$\hat{\varrho} \geq \sup_{\underline{x} \in R^n} \left| < \frac{\underline{x}}{\left(\int_{t_0}^{t_1} \|W^*(t_1,s)\underline{x}\|^q ds\right)^{\frac{1}{q}}}, \underline{x}_1 - V(t_1)x_0 > \right| =$$

$$\sup_{\left(\int_{t_0}^{t_1} \|W^*(t_1,s)\underline{x}\|^q ds\right)^{\frac{1}{q}} = 1} < \underline{x}, \underline{x}_1 - V(t_1)\underline{x}_0 > = \varrho' \qquad \text{II(13)}$$

Therefore we need only to show that $\hat{\varrho}$ cannot be larger than ϱ'. Let us suppose the contrary. Then there exists a positive constant σ such that $\varrho' < \sigma < \hat{\varrho}$.

Since $\sigma < \hat{\varrho}$, by II(10), the system is not ($\underline{x}_0, \underline{x}_1, K, U_p^{\sigma}$) controllable, and there exists a vector $\underline{x}' \in R^{n'}$

$$|< \underline{x}', \underline{x}_1 - V(t_1)\underline{x}_0 >| > \sigma\left(\int_{t_0}^{t_1} \|W^*(t_1,s)x'\|^q ds\right)^{\frac{1}{q}}. \qquad \text{II(14)}$$

Since $\sigma > 0$, $|< \underline{x}', \underline{x}_1 - V(t_1)\underline{x}_0 >|$ is positive and therefore, since $\hat{\underline{u}} \in \bar{U}$, hence $\hat{\varrho} > 0$, by II(11) also $\left(\int_{t_0}^{t_1} \|W^*(t_1,s)x'\|^q ds\right)^{\frac{1}{q}}$ is positive, and we may therefore consider the quantity

$$\underline{\eta}' = \frac{\underline{x}'}{\left(\int_{t_0}^{t_1} \|W^*(t_1,s)\underline{x}'\|^q ds\right)^{\frac{1}{q}}}$$

Obviously, $\left(\int_{t_0}^{t_1} \|W^*(t_1,s)\underline{\eta}'\|^q ds\right)^{\frac{1}{q}} = 1$

and by II(14) we have

$$|<\eta', \underline{x}_1 - V(t_0)\underline{x}_0>| > \sigma$$

and therefore

$$<\eta', \underline{x}_1 - V(t_1)\underline{x}_0> > \varrho'$$

which contraddicts II(13').

From the contraddiction, we have $\hat{\varrho} = \varrho'$ and the theorem is proved.

We prove now the following Corollary

Corollary II(2)

The supremum in II(12) is assumed

<u>Proof</u>

Take a sequence $\{\varrho_n\}$ with $\varrho_n < \hat{\varrho}$ and converging to $\hat{\varrho}$.

Then for each n there exists a vector $\underline{x}'_n \in R^n$, $\|\underline{x}'_n\| = 1$, such that

$$|<\underline{x}'_n, \underline{x}_1 - V(t_1)\underline{x}_0>| > \varrho_n \left(\int_{t_0}^{t_1} \|W^*(t_1,s)\, x_n\|^q ds\right)^{\frac{1}{q}}$$

Reasoning as in Theorem II,2, as n tends to infinity we have $\lim_{n\to\infty} \underline{x}'_n = \hat{\underline{x}}'$, with $\|\hat{\underline{x}}'\| = 1$,

$$|<\hat{\underline{x}}', \underline{x}_1 - V(t_1)\underline{x}_0>| \geq \hat{\varrho}\left(\int_{t_0}^{t_1} \|W^*(t_1,s)\,\hat{\underline{x}}'\|^q ds\right)^{\frac{1}{q}} \geq 0$$

Since the system is $x_0, x_1, \mathcal{U}_p^{\varrho}$, K controllable, we have

$$0 < \;<\hat{x}', \underline{x}_1 - V(t_1)\, x_0> \;\leq \hat{\varrho}\left(\int_{t_0}^{t_1} \|W^*(t_1,s)\hat{x}'\|^q ds\right)^{\frac{1}{q}}$$

and therefore we may write

$$\hat{\varrho} = \frac{|<\hat{x}', \underline{x}_1 - V(t_1)x_0>|}{\left(\int_{t_0}^{t_1} \| W^*(t_1; s)\hat{x}'\|^q ds\right)^{\frac{1}{q}}}$$

or

$$\hat{\varrho} = <\hat{\underline{x}}, \underline{x}_1 - V(t_1)x_0> = \max_{\left(\int_{t_0}^{t_1} \| W^*(t_1,s)x\|^q ds\right)^{\frac{1}{q}}} |<\underline{x}, \underline{x}_1 - V(t_1)x_0>| \qquad II(15)$$

and the corollary is proved

Going back to consider the time optimal problem, let us suppose that $\bar{K} = [t_0, \bar{t}]$ is the minimal time interval over which the system I(1) can be transferred from $\underline{x}_0$ at time t_0 to $\underline{x}_1$, with $\underline{u} \in \mathcal{U}_p^{\varrho}$, i.e. with an $\underline{u}$ such that $\|\underline{u}\| \leq \varrho$. So in the time interval $\bar{K} = [t_0, \bar{t}]$ there exist controls with norm not larger than ϱ which effect the desired transfer.

On the other hand, we know from theorem II,4 and Corollary II,2 that, for the interval $\bar{K}$ considered as given, the minimal norm controls have norm $\hat{\varrho}$:

$$\hat{\varrho} = \sup_{\left(\int_{t_0}^{\bar{t}} \| W^*(t_1,s)\underline{x}\|^q ds\right)^{\frac{1}{q}} = 1} |<\underline{x}, \underline{x}_1 - V(t_1)\underline{x}_0>| = |<\hat{x}, \underline{x}_1 - V(t_1)x_0>|$$

We may rise the following question: with reference to the same system, the same points to, ϱ and hence $\mathcal{U}_p^\varrho$, can $\hat{\varrho}$ be less than ϱ ? In other words, once we found some control which is time optimal relative to two points and to an upper bound for the norm of admissible controls, is its norm also minimal with respect to the norms of controls which effect the desired transfer in the minimal interval $\bar{K} = [t_0, \bar{t}]$?

We shall give a partial answer to this question in the following

Corollary II,3

If the system I(1) is controllable relative to any pair of points of R^n, to class $\mathcal{U}_p^\varrho$ of admissible controls and to the time interval $\bar{K} = [t_0, \bar{t}]$, where $\bar{K}$ is the minimal time interval over which the system is controllable relative to given $\underline{x}_0$, $\underline{x}_1$, and to $\mathcal{U}_p^\varrho$, with ϱ given, then $\varrho = \hat{\varrho}$, where $\hat{\varrho}$ is the minimal value of the norm of controls which effect the given transfer from $\underline{x}_0$ to $\underline{x}_1$, in $\bar{K}$.

Proof

Since $\bar{K} = [t_0, \bar{t}]$ is a time optimal interval

by Theorem II,2 there exists a vector $\underline{x}'$ such that II(6) holds.

Since the system is controllable relative to any $\underline{x}_0$, $\underline{x}_1$, to $\mathcal{U}_p^\varrho$ and to $\bar{K} = [t_0, \bar{t}]$, by the already quoted usual theorem (cfr. p.26 chapter I), the matrix

$$Y(t_0,\bar{t}) = \int_{t_0}^{\bar{t}} W(\bar{t},s)\, W^*(\bar{t},s)\, ds$$ is positive definite and therefore the vector $W^*(\bar{t},s)\,\underline{x}'$ cannot be a. e. zero in $\bar{K}$ and the integral $\left(\int_{t_0}^{\bar{t}} \| W^*(\bar{t},s)\,\underline{x}'\|^q ds\right)^{1/q}$ is positive. We may therefore write

$$\hat{\varrho} = \frac{|<\underline{x}', \underline{x}_1 - V(\bar{t})\underline{x}_0>|}{\left(\int_{t_0}^{\bar{t}} \| W^*(\bar{t},s)\,\underline{x}'\|^q ds\right)^{1/q}} = |<\hat{x}, \underline{x}_1 - V(t_1)x_0>|$$

Since, by the definition II(10), $\hat{\varrho}$ is the minimal norm of controls belonging to $\mathcal{U}_p^\varrho$ which transfer $\underline{x}_0$ into $\underline{x}_1$ within $\bar{K} = [t_0, \bar{t}]$, then $\hat{\varrho} \leq \varrho$. By II(12) or II(15) we have $\hat{\varrho} \geq \varrho$, and therefore $\hat{\varrho}$ is equal to ϱ, and the corollary is proved.

In any case, given $\underline{x}_0$ and $\underline{x}_1$, and a class of admissible controls, if $\bar{K} = [t_0, \bar{t}]$ is the minimal time interval in which the system I(1) is controllable, there exists a control $\hat{\underline{u}}$ which transfers $\underline{x}_0$ into $\underline{x}_1$, in $\bar{K}$ and with minimal norm $\hat{\varrho}$. In what follows, we shall always refer to the system I(1), to given points $\underline{x}_0$ and $\underline{x}_1$, to given initial time t_0 and time interval $\bar{K} = [t_0, \bar{t}]$. This time interval may be considered either the minimal time interval for transferring $\underline{x}_0$ into $\underline{x}_1$ with controls $\underline{u} \in \mathcal{U}_p^{\varrho}$ (ϱ given, if the problem is of time minimization) or simply the given time interval (if the problem is of norm minimization).

With these conventions, and using all results we got until now, we are now going to deduce the "optimal" control functions, which are to be intended as both time and norm minimal in the first case and as norm minimal in the second. In both cases, the norm of the optimal control is given by II(15).

We want now to find the optimal control function $\hat{\underline{u}}$ (optimal in the sense just specified).

We shall refer to the general norm in R^m:

$$\| \underline{u}(t) \|_r = \left\{ \sum_1^m {}_i \, | \underline{u}(t) |_i^r \right\}^{\frac{1}{r}} \qquad 1 \leq r \leq \infty \qquad \text{II(16)}$$

(notice that we introduced the index r in $\| u(t) \|_r$ for clarity)

We shall use from now on the notation $\left| \| \underline{u}(\cdot) \|_r \right|_p$ for the following general norm.

$$\left| \| \underline{u}(\cdot) \|_r \right|_p = \left(\int_{t_0}^{\bar{t}} \| \underline{u}(s) \|_r^p \, ds \right)^{\frac{1}{p}} \qquad 1 < p \leq \infty$$

As we already mentioned, all results we got until now still hold with this general norm, of course with the norm of $W^*(\bar{t},t)\underline{x}$ taken in conjugate space of the one for $\underline{u}(t)$, so with the new notation we have

$$\| W^*(\bar{t},t)\,\hat{\underline{x}}\,\|_s = \Big\{ \sum_{1}^{m} {}_i \,| W^*(\bar{t},t)\,\underline{x}|_i^s \Big\}^{\frac{1}{s}} \qquad \Big(\frac{1}{r}+\frac{1}{s}=1\Big)$$

Since

$$\hat{\varrho} = \Big| \| \hat{\underline{u}}(\cdot)\|_r \Big|_p \tag{II(18)}$$

we have that there exists a vector $\hat{\underline{x}}$ such that

$$|<\hat{\underline{x}}\,,\underline{x}_1 - V(\bar{t})\,\underline{x}_0>| = \hat{\varrho}\,\Big(\int_{t_0}^{\bar{t}} \| W^*(\bar{t},s)\,\hat{\underline{x}}\,\|_s^q\, ds\Big)^{\frac{1}{q}} > 0 \quad \Big(\frac{1}{r}+\frac{1}{q}=1\Big) \tag{II(19)}$$

Since $\hat{\underline{u}}$ transfers $\underline{x}_0$ into $\underline{x}_1$ in the interval $\bar{K} = [t_0,\bar{t}]$, we have

$$\underline{x}_1 - V(\bar{t})\,\underline{x}_0 = \int_{t_0}^{\bar{t}} W(\bar{t},s)\,\hat{\underline{u}}(s)\,ds \tag{II(20)}$$

where, as usual, $W(\bar{t},t) = V(\bar{t})\,V^{-1}(t)B(t)$.

Putting together II(19) and II(20) and using again the adjoint $W^*(\bar{t},t)$ of $W(\bar{t},t)$, we have

$$\hat{\varrho}\,\Big(\int_{t_0}^{t} \| W^*(\bar{t},s)\,\hat{\underline{x}}\,\|_s^q\, ds\Big)^{\frac{1}{q}} = \Big| \int_{t_0}^{\bar{t}} < W^*(\bar{t},s)\,\hat{\underline{x}}\,,\,\hat{\underline{u}}(s)>\, ds \Big|$$

Using Hölder's inequality in the finite dimensional and infinite dimensional case we may write

$$\hat{\varrho}\left(\int_{t_0}^{\bar{t}}\|W^*(\bar{t},s)\,\hat{\underline{x}}\|_s^q\right)^{\frac{1}{q}} ds = \left|\int_{t_0}^{\bar{t}} < W^*(\bar{t},s)\,\hat{\underline{x}},\hat{\underline{u}}(s) > ds\right| \leq$$

$$\int_{t_0}^{\bar{t}} \left| < W^*(\bar{t},s)\,\hat{\underline{x}},\hat{\underline{u}}(s) > \right| ds \leq \int_{t_0}^{\bar{t}} \|W^*(\bar{t},s)\,\hat{\underline{x}}\|_s \, \|\hat{\underline{u}}(s)\|_r \, ds \leq$$

$$\left(\int_{t_0}^{\bar{t}}\|\hat{\underline{u}}(s)\|_r^p\, ds\right)^{\frac{1}{p}}\left(\int_{t_0}^{\bar{t}}\|W^*(\bar{t},s)\,\hat{\underline{x}}\|_s^q\, ds\right)^{\frac{1}{q}} = \hat{\varrho}\left(\int_{t_0}^{\bar{t}}\|W^*(\bar{t},s)\,\hat{\underline{x}}\|_s^q\, ds\right)^{\frac{1}{q}}$$

II(21)

Since in II(21) the first and last member are equal, we must have equality at each step. Recalling the conditions for Hölder's inequality to hold as an equality, we have a.e. in $\bar{K}$:

$$\|\hat{\underline{u}}(t)\|_r = C\,\|W^*(\bar{t},t)\,\hat{\underline{x}}\|_s^{\frac{q}{p}} \qquad \text{II(22)}$$

$$\hat{u}_i(t) = K(t)\left[W^*(\bar{t},t)\,\hat{\underline{x}}\right]_i^{\frac{s}{r}} \qquad \text{II(23)}$$

and

$$\text{sign } u_i(t) = \text{sign}\left[W^*(\bar{t},t)\,\hat{\underline{x}}\right]_i \qquad \text{II(24)}$$

From II(22) and II(18) we may determine the constant C . Indeed

$$\hat{\varrho} = \left|\,\|\hat{\underline{u}}(\cdot)\|_r\right|_p = C\left(\int_{t_0}^{\bar{t}}\|W^*(\bar{t},s)\,\hat{\underline{x}}\|_s^{\frac{q}{p}\cdot p}\, ds\right)^{\frac{1}{p}} = C\left|\,\|W^*(\bar{t},t)\,\hat{\underline{x}}\|_s\right|_q^{\frac{q}{p}}$$

hence

$$C = \frac{\hat{\varrho}}{\left| \| W^*(\bar{t},t)\hat{\underline{x}} \|_s \right|_q^{\frac{q}{p}}} = \frac{\hat{\varrho}}{\left| \| W^*(\bar{t},t)\hat{\underline{x}} \|_s \right|_q^{q-1}} \qquad \text{II(25)}$$

Since (cfr. II(12) or II(15)) the vector $\hat{\underline{x}}$ is determined up to a moltiplicative real constant, and since we suppose as always $\underline{x}_1 - V(\bar{t})\underline{x}_0 \neq 0$, we may determine $\hat{\underline{x}}$ such that the scalar product $|< \hat{\underline{x}} , \underline{x}_1 - V(\bar{t})\, \underline{x}_0 >|$ is equal to one. In correspondence to such $\hat{\underline{x}}$ we have from II(15) $\hat{\varrho} = \frac{1}{\left| \| W^*(\bar{t},t)\hat{\underline{x}} \|_s \right|_q}$ and therefore

$$C = \frac{1}{\left| \| W^*(\bar{t},t)\, \hat{\underline{x}} \|_s \right|_q^{q}} \qquad \text{II(26)}$$

To determine the function $K(t)$ let us take the finite dimensional norm of $\hat{\underline{u}}\ (t)$, using II(23)

$$\| \underline{u}(t) \|_r = |K(t)| \left[\sum_1^m{}_i \left| W^*(\bar{t},t)\hat{\underline{x}} \right|_i^{\frac{s}{r}\cdot r} \right]^{\frac{1}{r}} = |K(t)|\, \| W^*(\bar{t},t)\, \hat{\underline{x}} \|^{\frac{1}{r}} \qquad \text{II(27)}$$

From II(22) and II(26), II(27) we have

$$|K(t)| = \frac{1}{\left| \| W^*(\bar{t},t)\hat{\underline{x}} \|_s \right|_q^{q}} \quad \| W^*(\bar{t},t)\, \hat{\underline{x}} \|_s^{\left(\frac{q}{p} - \frac{s}{r}\right)} \qquad \text{II(28)}$$

From II(23), II(24), II(28) we eventually have

$$\hat{u}_i(t) = \frac{1}{\left|\|W^*(\bar{t},t)\hat{\underline{x}}\|_s\right|_q^q} \|W^*(\bar{t},t)\hat{\underline{x}}\|_s^{\left(\frac{q}{p}-\frac{s}{r}\right)} \, |W^*(\bar{t},t)\hat{\underline{x}}|_i^{s-1} \operatorname{sign}\left(W^*(\bar{t},t)\hat{\underline{x}}\right)_i \qquad \text{II(29)}$$

which is the general expression for the optimal control when the norm of $\underline{u}$ is the general norm

$$\left|\|\underline{u}(\cdot)\|_r\right|_p = \left(\int_{t_0}^{\bar{t}} \left\{\sum_1^m {}_i \, |u_i(s)|^r\right\}^{\frac{p}{r}} ds\right)^{\frac{1}{p}}$$

$\frac{1}{r} + \frac{1}{s} = 1$, $\frac{1}{p} + \frac{1}{q} = 1$, and $\hat{\underline{x}}$ is given by II(19) with the additional constraint $|<\hat{\underline{x}}, \underline{x}_1 - V(\bar{t})\underline{x}_0>| = 1.$

Many particular cases of the general formula II(29) are interesting. For example if $\underline{u} \in \mathcal{L}_{2,2}$ and its norm becomes

$$\left|\|\underline{u}(\cdot)\|_2\right|_2 = \left(\int_{t_0}^{\bar{t}} \sum_1^m {}_i \, |u_i(s)|^2 ds\right)^{\frac{1}{2}}$$

so as to have the intuitive meaning of "energy", the optimal control function is the following:

$$\hat{\underline{u}}_i(t) = \frac{1}{\int_{t_0}^{\bar{t}} \sum_1^m {}_i \, |W^*(\bar{t},s)\hat{\underline{x}}|_i^2 ds} \left(W^*(\bar{t},t)\hat{\underline{x}}\right)_i \; ;$$

if $\underline{u} \in \mathcal{L}_{\infty,\infty}$ and its norm is

$$\left|\|\underline{u}(\cdot)\|_\infty\right|_\infty = \operatorname*{ess\,sup}_{t \in \bar{K}} \left[\max_{i=1\ldots m} |\underline{u}_i(t)|\right]$$

the optimal control function is

$$\hat{\underline{u}}_i(t) = \frac{1}{\int_{t_0}^{\bar{t}} \sum_1^m {}_i \left| W^*(\bar{t},s)\hat{x} \right|_i ds} \operatorname{sign} \left[W^*(\bar{t},t)\hat{x} \right]_i$$

In general, the type of norm to be chosen is given by physical considerations, as it is obvious.

Let us consider the case $u \in \mathcal{L}_{r,\infty}$, or

$$\left| \| u(\cdot) \|_r \right|_\infty = \text{ess sup} \left\{ \sum_1^m {}_i |u_i(t)|^r \right\}^{\frac{1}{r}} ; \qquad (1 \leq r \leq \infty)$$

and let us see how is now possible to derive the well-known Pontryagin's maximum principle for this case.

Let us take the system I(1) with $\underline{u} \equiv 0$:

$$\dot{\underline{x}} = A(t)\, \underline{x} \qquad \text{II(29)}$$

and its adjoint equation

$$\dot{\underline{\Psi}} = -A^*(t)\, \underline{\Psi} \qquad \text{II(30)}$$

If we let the fundamental matrix solution $V(t)$ of II(29) be non singular at $\bar{t}$, then is non-singular, hence invertible, almost everywhere.

Calling $\Psi(t)$ the fundament matrix solution of II(30)

with $\Psi^*(\bar{t}) = V^{-1}(\bar{t})$, it is immediate to see that

$$\Psi^*(t) = V^{-1}(t)$$

(Indeed, $\frac{d}{dt}\left[\Psi^*(t)V(t)\right] = -\left[A^*(t)\Psi(t)\right]^* V(t) + \Psi^*(t)A(t)V(t) = 0$, hence $\Psi^*(t)\,V(t) = I$, where I is the identity matrix).

As a consequence, we may write the general vector solution of II(30) in the form

$$\underline{w}(t) = \Psi(t)\,\Psi^{-1}(\bar{t})\,\hat{\underline{x}} = V^{-1*}(t)\,V^*(\bar{t})\,\hat{\underline{x}} \qquad \text{II(31)}$$

where $\hat{\underline{x}} = \underline{w}(\bar{t})$.

The optimal control law $\hat{\underline{u}}$ makes all inequalities of II(21) be equalities, as we saw. In particular,

$$|<W^*(\bar{t},t)\hat{\underline{x}}, \hat{\underline{u}}(t)>| = \|W^*(\bar{t},t)\hat{\underline{x}}\|_s \, \|\hat{\underline{u}}(t)\|_r \qquad \text{a.e. in } \bar{K}$$

or equivalently:

$$|<V^{-1*}(t)V^*(\bar{t})\hat{\underline{x}}, B(t)\hat{\underline{u}}(t)>| = |<B^*(t)V^{-1*}(t)V^*(\bar{t})\hat{\underline{x}}, \hat{\underline{u}}(t)>| =$$

$$= \|B^*(t)V^{-1*}(t)V^*(\bar{t})\hat{\underline{x}}\|_s \, \|\hat{\underline{u}}(t)\|_r$$

and since for every $\underline{u}$ such that $\operatorname*{ess\,sup}_{t \in \bar{K}} \left(\sum_{1}^{m}{}_i |u_i(t)|^r\right)^{\frac{1}{r}} \leq \varrho$

$$|<V^{-1*}(t)V^*(\bar{t})\underline{x}, B(t)\underline{u}(t)>| \leq \|B^*(t)V^{-1*}(t)V^*(\bar{t})\hat{\underline{x}}\|_s \, \|\underline{u}(t)\|_r \, , \qquad \text{II(32)}$$

we may conclude that for every t the function $\hat{\underline{u}}(t)$ makes the quantity $\left| < V^{-1*}(t) V^{*}(\bar{t}) \underline{x}, B(t)\underline{u}(t)> \right|$ maximal with respect to all $\underline{u}(t)$ such that $\left| \| u(\cdot) \|_r \right|_\infty \leq \hat{\varrho}$, i.e. to all admissible controls. Since, by II(31), $\underline{w}(t) = V^{-1*}(t) V^{*}(\bar{t}) \hat{\underline{x}}$ is the solution of the adjoint equation $\dot{\underline{w}}(t) = -A^{*}(t) \underline{w}(t)$ satisfying $\underline{w}(\bar{t}) = \hat{\underline{x}}$, we may say that the optimal control $\hat{\underline{u}}(t)$ maximazes for every t a.e. and with respect to all admissible controls, the scalar product $\left| < \underline{w}(t), B(t) \underline{u}(t) > \right|$, where $\underline{w}(t)$ is an appropriate solution of the quoted adjoint vector equation. This is precisely Pontryagin's maximum principle for this case.

Let us now go back to the general optimal control II(29) and notice that the only quantity whose computation is not explicit in it is the vector $\hat{\underline{x}}$, which corresponds to

$$\max_{|<\underline{x},\underline{x}_1 - V(\bar{t})\underline{x}_0>| = 1} \frac{1}{\left(\int_{t_0}^{\bar{t}} \| W^{*}(\bar{t},s)\, \underline{x} \|_s^q \, ds \right)^{\frac{1}{q}}} \qquad \left(\frac{1}{p} + \frac{1}{q} = 1 ;\ \frac{1}{r} + \frac{1}{s} = 1 \right)$$

For a general type of norm, this computation is a rather difficult problem (cfr.for example [5] , [6]), which we are not going to treat here.

We shall only show how the computation of $\hat{\underline{x}}$ becomes an easy task when $\underline{u} \in L_{2,2}$, hence $\underline{u}$ belongs to an Hilbert space From II (29) we see that in this case the form of the optimal control is

$$\underline{\hat{u}}(t) = \frac{W^*(\bar{t},t)\,\hat{\underline{x}}}{\int_{t_0}^{\bar{t}} \langle W^*(\bar{t},s)\hat{\underline{x}}, W^*(\bar{t},s)\hat{\underline{x}}\rangle ds} = \hat{\varrho}^2\, W^*(\bar{t},t)\,\hat{\underline{x}}. \qquad \text{II(33)}$$

Since $\underline{\hat{u}}$ transfers $\underline{x}_0$ into $\underline{x}_1$ at $\bar{t}$, we have, putting again $\underline{z}(\bar{t}) = \underline{x}_1 - V(\bar{t})\,\underline{x}_0$,

$$\int_{t_0}^{\bar{t}} W(\bar{t},s)\,\underline{\hat{u}}(s)\,ds = \underline{z}(\bar{t}).$$

which, by II(33) becomes

$$\hat{\varrho}^2 \int_{t_0}^{\bar{t}} W(\bar{t},s)\, W^*(\bar{t},s)\,\hat{\underline{x}}\, ds = \underline{z}(\bar{t})$$

or, with the usual position $Y(\bar{t}) = \int_{t_0}^{\bar{t}} W(\bar{t},s)\, W^*(\bar{t},s)\, ds$

$$\hat{\varrho}^2\, Y(\bar{t})\,\hat{\underline{x}} = \underline{z}(\bar{t}). \qquad \text{II(34)}$$

and recalling

$$\left| \langle \hat{\underline{x}}, \underline{z}(\bar{t}) \rangle \right| = 1 \qquad \text{II(35)}$$

the determination of $\hat{\underline{x}}$ is reduced in this case to the solution of the equation II(34) with constraint II(35).

Let us distinguish two cases: in the first one the system I(1) is controllable at time $\bar{t}$ relative to <u>any</u> pair of points of R^n. In this case $Y(\bar{t})$ is non singular, therefore invertible, and the equation II(34) has the unique solution

$$\hat{\underline{x}} = \frac{1}{\hat{\varrho}^2} Y^{-1}(\bar{t})\, \underline{z}(\bar{t}) \qquad \text{II(36)}$$

and form II(33) the expression of the optimal control law $\hat{\underline{u}}$ is

$$\hat{\underline{u}}(t) = B^*(t)\, V^{-1*}(t)\, V^*(\bar{t})\, Y^{-1}(\bar{t})\, \underline{z}(\bar{t}) \qquad \text{for } t \in [t_0, \bar{t}] \qquad \text{II(37)}$$

In the second case, the matrix $Y(\bar{t})$ is singular, but the system is still controllable with respect to $\underline{x}_0$, $\underline{x}_1$, $K = [t_0, \bar{t}]$, $\hat{\varrho}$, and therefore the matrix $C(t) = Y(\bar{t}) - \frac{1}{\hat{\varrho}^2}\, \underline{z}(\bar{t})$ is positive semidefinite. We shall see that it is still possible to find an explicit expression of the optimal control law. To do that, we need the following

Theorem II,5

Let $R[Y(\bar{t})]$ be the range of the matrix $Y(\bar{t})$. Then, if $C(\bar{t})$ is positive semidefinite, the vector $\underline{z}(\bar{t})$ belongs to $R[Y(\bar{t})]$.

Proof

Let $N[Y(\bar{t})]$ be the null space of $Y(\bar{t})$, i.e. the set of vectors $\underline{x}$ such that $Y(\bar{t})\,\underline{x} = 0$. (The dimension of the null space $N[Y(\bar{t})]$ is equal to the multeplicity of 0 as an eigenvalue of $Y(\bar{t})$). The, since $Y(\bar{t})$ is real and symmetric, its eigenvectors span the space R^n , and

$$R^n = R\left[Y(\bar{t})\right] \oplus N\left[Y(\bar{t})\right]$$

Let P_N be the projection operator of R^n into $N[Y(\bar{t})]$, and consider the vector $P_N \underline{z}(\bar{t})$. Since $C(\bar{t})$ is positive semi definite, we have

$$< C(\bar{t}) P_N \underline{z}(\bar{t}), P_N \underline{z}(\bar{t}) > \ \geq 0 \qquad \text{II(38)}$$

and, by definition of $N\left[Y(\bar{t})\right]$,

$$< C(\bar{t}) P_N \underline{z}(\bar{t}), P_N \underline{z}(\bar{t}) > = < Y(\bar{t}) P_N \underline{z}(\bar{t}), \underline{z}(\bar{t}) > - \qquad \text{II(39)}$$

$$-\frac{1}{\hat{\varrho}^2} < P_N \underline{z}(\bar{t}), \underline{z}(\bar{t}) > = -\frac{1}{\hat{\varrho}^2}\left(< P_N \underline{z}(\bar{t}), \underline{z}(\bar{t}) >\right) \leq 0$$

From II(38) and II(39) we have $< P_N \underline{z}(\bar{t}), z(\bar{t}) > = 0$ and therefore $P_N \underline{z}(\bar{t})$ is orthogonal to $\underline{z}(\bar{t})$, which means that $\underline{z}(\bar{t})$ does not have components in $N\left[Y(\bar{t})\right]$ and belongs to $M\left[Y(\bar{t})\right]$, as we wanted to prove.

Let now ℓ be the dimension of $N\left[Y(\bar{t})\right]$, so that $n-\ell$ is the dimension of $M\left[Y(\bar{t})\right]$. The vector $\underline{z}(\bar{t})$ can then be written as $\underline{z}(\bar{t}) = \sum_{\ell+1}^{n}{}_i\, \zeta_i\, \underline{d}_i$, where $\underline{d}_i$ are the eigenvectors of $Y(\bar{t})$ corresponding to the non-null eigenvalues of $Y(\bar{t})$. Since $\hat{\underline{x}}$ can be written as $\hat{\underline{x}} = \sum_{1}^{n}{}_i\, \xi_i\, \underline{d}_i$, where $\underline{d}_1 \ldots \underline{d}_\ell$ are the vectors spanning $N\left[Y(\bar{t})\right]$, equa tion II(34) becomes

$$\hat{\varrho}^2 Y(\bar{t}) \sum_1^n {}_i \zeta_i \underline{d}_i = \sum_{l+1}^n {}_i \mathfrak{z}_i \underline{d}_i$$

or

$$\hat{\varrho}^2 \sum_{l+1}^n {}_i \zeta_i \lambda_i \underline{d}_i = \sum_{l+1}^n {}_i \mathfrak{z}_i \underline{d}_i$$

where λ_i are the non-null (and real, since $Y(\bar{t})$ is real and symmetric) eigenvalues of $Y(\bar{t})$. The solution of II(34) is therefore

$$\hat{\underline{x}} = \frac{1}{\hat{\varrho}^2}\left(\sum_1^l {}_i \zeta_i \underline{d}_i + \sum_{l+1}^n {}_i \frac{\mathfrak{z}_i}{\lambda_i} \underline{d}_i\right) \qquad \text{II(40)}$$

and from II(33) we see that the optimal control law is

$$\hat{\underline{u}}(t) = B^*(\bar{t}) V^{-1*}(t) V^*(\bar{t}) \left(\sum_1^l {}_i \zeta_i \underline{d} + \sum_{l+1}^n {}_i \frac{\mathfrak{z}_i}{\lambda_i} \underline{d}_i\right) \qquad \text{II(41)}$$

where $\zeta_1 \dots \zeta_l$ are arbitrary constants.

The expression II(40) is of course valid also if $Y(\bar{t})$ is non-singular, in which case $l = 0$ and II(40) is equivalent to II(37), with $\hat{\underline{x}} = Y^{-1}(\bar{t}) \underline{z}(\bar{t}) = \sum_1^n {}_i \frac{\mathfrak{z}_i}{\lambda_i}$, where $\lambda_1 \dots \lambda_n$ are the (positive) eigenvalues of $Y(\bar{t})$ and $\mathfrak{z}_1 \dots \mathfrak{z}_n$ the components of $\underline{z}(\bar{t})$ along the eigenvectors $\underline{d}_1 \dots \underline{d}_n$ of $Y(\bar{t})$.

Let us now notice from II(40) that if 0 is an eigenvalue of $Y(\bar{t})$ with multiplicity l , there are ∞^l op timal control laws, all of them transferring $\underline{x}_0$ in $\underline{x}_1$ with controls which have the same norm $\hat{\varrho}$.

We may check that the value of $\xi_1 \ldots \zeta_n$ does not affect the norm $\hat{\varrho}$ simply by computing $\hat{\varrho}$; from II(35) and II(40) we have

$$1 = \left| < \hat{\underline{x}}, \underline{z}(\bar{t}) > \right| = \frac{1}{\hat{\varrho}^2} \left| < \sum_1^{\ell}{}_i \zeta_i \underline{d}_i + \sum_{\ell+1}^{n}{}_i \frac{z_i}{\lambda_i} \underline{d}_i \, , \sum_{\ell+1}^{n}{}_i z_i \underline{d}_i > \right| = \frac{1}{\hat{\varrho}^2} \sum_{\ell+1}^{n}{}_i \frac{z_i^2}{\lambda_i}$$

from which

$$\hat{\varrho} = \left(\sum_{\ell+1}^{n}{}_i \frac{z_i^2}{\lambda_i} \right)^{\frac{1}{2}} \qquad \text{II(42)}$$

The formula II(42) with $\ell = 0$ is of course valid also when $Y(\bar{t})$ is non-singular. In this case, from II(35) and II(36) we have $\frac{1}{\hat{\varrho}^2} \left| < Y^{-1}(\bar{t}) \underline{z}(\bar{t}) , \underline{z}(\bar{t}) > \right| = 1$, hence

$$\hat{\varrho} = \left| < Y^{-1}(\bar{t}) \underline{z}(\bar{t}) , \underline{z}(\bar{t}) > \right|^{\frac{1}{2}} .$$

E x e r c i s e

Given the system

$$\underline{\dot{x}} = -\frac{1}{4}\begin{Vmatrix} 2 & 20 \\ 5 & 2 \end{Vmatrix}\underline{x} + \begin{Vmatrix} 2 & 4 \\ 1 & 2 \end{Vmatrix}\underline{u}(t) \qquad \underline{x}_0 = \begin{vmatrix} 2 \\ -3 \end{vmatrix} \quad \underline{x}_1 = \begin{vmatrix} 8 \\ -4 \end{vmatrix} \qquad E(1)$$

and admissible controls $\underline{u}$ such that

$$\int_0^T \|\underline{u}(s)\|^2 ds \leq 3$$

where $\|\underline{u}(t)\|$ is the euclidean norm of $\underline{u}(t)$,

a) prove that it is controllable relative to $\underline{x}_0$, $\underline{x}_1$ to the interval $K = [0, T]$ for some $T > 0$, and to the given class of admissible controls.

b) determine the minimal time $\bar{t}$ such that the system is controllable relative to the same quantities and construct a corresponding time optimal control.

c) verify whether there exists only one time optimal control

d) verify whether there exist admissible controls which transfer $\underline{x}_0$ in $\underline{x}_1$, and which are time optimal but not norm minimal.

e) if the class of admissible controls is $\mathcal{U} = \{\underline{u} : \sup_{t\in[0,T]} \|\underline{u}(t)\|\}$, with respect to which T is the system controllable?

S o l u t i o n

The eigenvalues of the matrix $A = -\frac{1}{4}\begin{Vmatrix} 2 & 20 \\ 5 & 2 \end{Vmatrix}$ are $\lambda_1 = 2$, $\lambda_2 = -3$.

The diagonal matrix $A^1 = S^{-1} A S$, with $S = \frac{1}{4}\begin{Vmatrix} 2 & 2 \\ -1 & 1 \end{Vmatrix}$ and $S^{-1} = \begin{Vmatrix} 1 & -2 \\ 1 & 2 \end{Vmatrix}$ is therefore $A^1 = \begin{Vmatrix} 2 & 0 \\ 0 & -3 \end{Vmatrix}$, and the fundamental matrix solution $V(t)$ of E(1) with $V(0) = I$ is

$$V(t) = \frac{1}{4}\begin{Vmatrix} 2 & 2 \\ -1 & 1 \end{Vmatrix}\begin{Vmatrix} e^{2t} & 0 \\ 0 & e^{-3t} \end{Vmatrix}\begin{Vmatrix} 1 & -2 \\ 1 & 2 \end{Vmatrix} .$$

Therefore

$$W(T,t) = W(T-t) = V(T)V^{-1}(t)B(t) = \frac{1}{4}\begin{Vmatrix} 2 & 2 \\ -1 & 1 \end{Vmatrix}\begin{Vmatrix} e^{2(T-t)} & 0 \\ -1 & e^{-3(T-t)} \end{Vmatrix}\begin{Vmatrix} 1 & -2 \\ 1 & 2 \end{Vmatrix}\begin{Vmatrix} 2 & 4 \\ 1 & 2 \end{Vmatrix} =$$

$$= e^{-3(T-t)}\begin{Vmatrix} 2 & 2 \\ -1 & 1 \end{Vmatrix}\begin{Vmatrix} 0 & 0 \\ 1 & 2 \end{Vmatrix} , \qquad \text{E(2)}$$

and the matrix product $W(T,t)\,W^*(T,t)$ is

$$W(T-t)W^*(T-t) = e^{-6(T-t)}\begin{Vmatrix} 2 & 2 \\ -1 & 1 \end{Vmatrix}\begin{Vmatrix} 0 & 0 \\ 0 & 5 \end{Vmatrix}\begin{Vmatrix} 2 & -1 \\ 2 & 1 \end{Vmatrix} \qquad \text{E(3)}$$

and therefore

$$Y(T) = \int_0^T W(T-s)\,W^*(T-s)\,ds = \frac{1-e^{-6T}}{6}\begin{Vmatrix} 2 & 2 \\ -1 & 1 \end{Vmatrix}\begin{Vmatrix} 0 & 0 \\ 0 & 5 \end{Vmatrix}\begin{Vmatrix} 2 & -1 \\ 2 & 1 \end{Vmatrix} =$$

$$5\,\frac{1-e^{-6T}}{6}\begin{Vmatrix}2 & 2\\ -1 & 1\end{Vmatrix}\begin{Vmatrix}0 & 0\\ 0 & 1\end{Vmatrix}\begin{Vmatrix}2 & -1\\ 2 & 1\end{Vmatrix} \qquad E(4)$$

We notice that since $Y(T)$ has 0 as eigenvalue it is not positive definite, hence the system is not controllable relative to <u>any</u> pair of points $\underline{x}_0, \underline{x}_1$.

From the previous position $\underline{z}(T) = \underline{x}_1 - V(T)\underline{x}_0$ we have

$$\underline{z}(T) = \begin{vmatrix}8\\ -4\end{vmatrix} - \frac{1}{4}\begin{Vmatrix}2 & 2\\ -1 & 1\end{Vmatrix}\begin{Vmatrix}e^{2T} & 0\\ 0 & e^{-3T}\end{Vmatrix}\begin{Vmatrix}1 & -2\\ 1 & 2\end{Vmatrix}\begin{vmatrix}2\\ -3\end{vmatrix} =$$

$$= \frac{1}{4}\begin{Vmatrix}2 & 2\\ -1 & 1\end{Vmatrix}\left[\begin{Vmatrix}1 & -2\\ 1 & 2\end{Vmatrix}\begin{vmatrix}8\\ -4\end{vmatrix} - \begin{Vmatrix}e^{2T} & 0\\ 0 & e^{-3T}\end{Vmatrix}\begin{vmatrix}8\\ -4\end{vmatrix}\right] \qquad E(5)$$

and therefore

$$\underline{z}(T) = \underline{z}(T)\underline{z}^*(T) = \begin{Vmatrix}2 & 2\\ -1 & 1\end{Vmatrix}\begin{Vmatrix}1-e^{2T} & -2\\ 1 & 2-e^{-3T}\end{Vmatrix}\begin{vmatrix}2\\ -1\end{vmatrix}\cdot$$

$$\cdot\begin{vmatrix}2 & -1\end{vmatrix}\begin{Vmatrix}1-e^{2T} & 1\\ -2 & 2-e^{-3T}\end{Vmatrix}\begin{Vmatrix}2 & -1\\ 2 & 1\end{Vmatrix}$$

From the definition of $C(t)$ and from E(4), E(5) we have

$$C(T) = Y(T) - \frac{1}{\varrho^2}\,z(T) =$$

$$= \begin{Vmatrix}2 & 2\\ -1 & 1\end{Vmatrix}\left[\begin{Vmatrix}0 & 0\\ 0 & \frac{5}{6}\left(1-e^{-6T}\right)\end{Vmatrix} - \frac{1}{\varrho^2}\begin{Vmatrix}\frac{1}{16}\left(16-8e^{2T}\right)^2 & c_1(T)\\ c_2(T) & e^{-6T}\end{Vmatrix}\right]\begin{Vmatrix}2 & -1\\ 2 & 1\end{Vmatrix}$$

E(6)

where $c_1(T)$ and $c_2(T)$ are functions obtainable from E(4), but which are not relevant for our purposes.

From E(6) we see that $C(T)$ has the same eigenvalues as the matrix

$$R(T) = \left\| \begin{matrix} -\frac{1}{\varrho^2}\frac{1}{16}\left(16-8e^{2T}\right)^2 & -\frac{1}{\varrho^2}c_1(T) \\ -\frac{1}{\varrho^2}c_2(T) & \frac{5}{6}\left(1-e^{-6T}\right)-\frac{1}{\varrho^2}e^{-6T} \end{matrix} \right\|$$

and therefore $C(T)$ has the same character as $R(T)$ as far as positive or negative semifinitness.

By inspecting $R(T)$ we easily see that a necessary condition for it to be positive semidefinite is that $16-8e^{2T}$ be zero, and this is verified only for $T = \frac{1}{2}\ln 2 = t^*$. So this is the only candidate for the system to be controllable. It turns out that the system is controllable relative to the given $\underline{x}_0$, $\underline{x}_1$, $\varrho = \sqrt{3}$ and the time interval $[0, t^*]$. To prove that, let us first see that the equation

$$Y(t^*)\underline{x} = \underline{z}(t^*) \qquad E(7)$$

has a solution. From E(4), E(5), the equation E(7) with $t^* = \frac{1}{2}\ln 2$ is

$$\frac{35}{48}\left\| \begin{matrix} 2 & 2 \\ -1 & 1 \end{matrix} \right\| \left\| \begin{matrix} 0 & 0 \\ 0 & 1 \end{matrix} \right\| \left\| \begin{matrix} 2 & -1 \\ 2 & 1 \end{matrix} \right\| \hat{\underline{x}} = \left\| \begin{matrix} 2 & 2 \\ -1 & 1 \end{matrix} \right\| \left| \begin{matrix} 0 \\ 2^{-\frac{3}{2}} \end{matrix} \right| \qquad E(8)$$

Since $\left\| \begin{matrix} 2 & 2 \\ -1 & 1 \end{matrix} \right\|$ is non singular, we may premultiply both members of E(8) by its inverse, getting

$$\frac{35}{12} \left\| \begin{matrix} 0 & 0 \\ 0 & 1 \end{matrix} \right\| \left\| \begin{matrix} 2 & -1 \\ 2 & 1 \end{matrix} \right\| \underline{\hat{x}} = \left| \begin{matrix} 0 \\ 4 \cdot 2^{-\frac{3}{2}} \end{matrix} \right|$$

or,

$$(2\hat{x}_1 + \hat{x}_2) = \frac{12}{35} \cdot 4 \cdot 2^{-\frac{3}{2}} \qquad \text{E(9)}$$

The equation E(9) clearly has solution. From the text we know that any solution $\hat{x}$ of it is such that the control $\underline{\hat{u}}(t) = B^*(t)\, V^{-1*}(t)\, V(t^*)\, \underline{\hat{x}}$ takes $\underline{x}_0$ into $\underline{x}_1$, at time t^*. Actually, E(9) has infinitely many solutions, all with this property and all corresponding to norm optimal controls. Among these solutions let us take one of the form $\underline{\hat{x}} = \lambda \begin{bmatrix} 1 \\ 2 \end{bmatrix}$, so that from E(9) we have $\lambda = \frac{12}{35}\, 2^{-\frac{3}{2}}$.

The corresponding norm-minimal control is

$$\underline{\hat{u}}(t) = \frac{6}{35} \begin{bmatrix} 1 \\ 2 \end{bmatrix} e^{3t} \qquad 0 \leq t \leq \frac{1}{2} \ln 2 \qquad \text{E(10)}$$

and its squared norm is

$$\hat{\varrho}^2 = \left(\frac{6}{35}\right)^2 5 \int_0^{\frac{1}{2}\ln 2} e^{6s}\, ds = \frac{6}{35} \qquad \text{E(11)}$$

Since $\hat{\varrho} = \sqrt{\frac{6}{35}} < 3$, the control $\underline{\hat{u}}$ is admissible and we have proved that the given system is controllable relative to $\begin{bmatrix} 2 \\ -3 \end{bmatrix}$, $\begin{bmatrix} 8 \\ -4 \end{bmatrix}$, time interval $\left[0, \frac{1}{2} \ln 2\right]$ and $\varrho = \sqrt{3}$.

Of course, since $t^* = \frac{1}{2} \ln 2$ is the only time for which the system is controllable, then it is also the minimal time, and the control $\hat{\underline{u}}(t)$ given by E(10) is not only norm-minimal but also time optimal.

Since in E(11) we saw that $\hat{\varrho} < \varrho = \sqrt{3}$, we expect that, besides all infinitely many norm minimal controls which effect the same transfer in the same time, there are other controls which do the same job, being also admissible but not norm-minimal. Indeed, let us add to the norm-minimal control $\hat{\underline{u}}(t)$ given by E(10) any other control $\underline{u}^1(t)$ such that

$$\int_0^{\frac{1}{2}\ln 2} V\left(\frac{1}{2} \ln 2\right) V^{-1}(s) B(s)\, \underline{u}^1(s)\, ds = 0 \qquad \text{E(12)}$$

Taking $\hat{\underline{u}}(s) = \begin{bmatrix} K_1 \\ K_2 \end{bmatrix}$, with K_1, and K_2 constants, we have from E(2)

$$\int_0^{\frac{1}{2}\ln 2} V\left(\frac{1}{2} \ln 2\right) V^{-1}(s) B(s) \begin{bmatrix} K_1 \\ K_2 \end{bmatrix} ds = C \begin{Vmatrix} 2 & 4 \\ 1 & 2 \end{Vmatrix} \begin{vmatrix} K_1 \\ K_2 \end{vmatrix} = C \begin{bmatrix} 2K_1 + 4K_2 \\ K_1 - 2K_2 \end{bmatrix}$$

with C constant, and therefore all controls of the type $\underline{u}^1(t) = \begin{bmatrix} K_1 \\ -\frac{1}{2} K_1 \end{bmatrix}$ satisfy E(12). Therefore, all infinitely many controls $\underline{u}$ of the type $\underline{u}(t) = \hat{\underline{u}}(t) + \begin{bmatrix} K_1 \\ -\frac{1}{2} K_1 \end{bmatrix}$, with

$|K_1| > 0$ limited by the constraint $\int_0^{\frac{1}{2}\ln 2} \|\underline{u}(s)\| ds < 3$ are admissible and transfer $\underline{x}_0$ in $\underline{x}_1$ in the time interval $\left[0, \frac{1}{2}\ln 2\right]$, and are therefore time optimal but not norm minimal. We may observe that we would expect such a situation from the considerations we made in the text (cfr.Corollary II,3), since in this example $Y(t^*)$ is singular.

Passing now to the last question e), let us verify the general controllability condition I(11) in this specific case.

In order for the system to be controllable at time T we must have $|<\underline{x}, \underline{z}(T)>| \leq \sqrt{3}\int_0^T \|W^*(T,s)\,\underline{x}\| ds$ for every $\underline{x} \in R^n$.

From E(2), E(5) we must therefore have

$$\left|<\left\|\begin{matrix} 2 & -1 \\ 2 & 1 \end{matrix}\right\| \left|\begin{matrix} x_1 \\ x_2 \end{matrix}\right| \left|\begin{matrix} 4\cdot 2e^{2T} \\ e^{-3T} \end{matrix}\right|>\right| \leq \sqrt{3}\int_0^T e^{-3(T-s)} \left\| \left\|\begin{matrix} 0 & 1 \\ 0 & 2 \end{matrix}\right\| \left\|\begin{matrix} 2 & -1 \\ 2 & 1 \end{matrix}\right\| \left|\begin{matrix} x_1 \\ x_2 \end{matrix}\right| \right\| ds \quad \forall x_1, \forall x_2. \qquad E(13)$$

Since the matrix $\left\|\begin{matrix} 2 & -1 \\ 2 & 1 \end{matrix}\right\|$ is non singular, the vector $\left|\begin{matrix} y_1 \\ y_2 \end{matrix}\right| = \left\|\begin{matrix} 2 & -1 \\ 2 & 1 \end{matrix}\right\| \left|\begin{matrix} x_1 \\ x_2 \end{matrix}\right|$ span the space R^2 as well a as $\left|\begin{matrix} x_1 \\ x_2 \end{matrix}\right|$ and therefore we may write

$$\left|<\left|\begin{matrix} y_1 \\ y_2 \end{matrix}\right|, \left|\begin{matrix} 4-2e^{2T} \\ e^{-3T} \end{matrix}\right|>\right| \leq \sqrt{3}\int_0^T e^{-3(T-s)} \left\| \left\|\begin{matrix} 0 & 1 \\ 0 & 2 \end{matrix}\right\| \left|\begin{matrix} y_1 \\ y_2 \end{matrix}\right| \right\| ds \qquad \forall y_1, \forall y_2$$

or

$$\left|\left(4-2e^{2T}\right)y_1 + e^{-3T}y_2\right| \leq \frac{\sqrt{5}}{\sqrt{3}}\left(1-e^{-3T}\right)|y_2| \qquad \forall y_1, \forall y_2 \qquad E(14)$$

as a condition of controllability instead of E(13).

A necessary condition for E(14) to hold is that the coefficent of y_1 be zero, i.e. $T = \frac{1}{2} \ln 2$. For such a value of T , E(14) becomes

$$\frac{2^{-\frac{3}{2}}}{1-2^{-\frac{3}{2}}} \left| y_2 \right| \leq \sqrt{\frac{5}{3}} \left| y_2 \right| \qquad \forall y_2$$

which is satisfied.

Therefore we may conclude that the system E(1) is controllable with respect to $\underline{x}_0$, $\underline{x}_1$, class of admissible controls $\mathcal{U} = \left\{ \underline{u} : \sup_{t \in [0, \frac{1}{2}\ln 2]} \| \underline{u}(t) \| \right\}$, only for the time interval $K = \left[0, \frac{1}{2} \ln 2 \right]$.

chapter 3

CONTROLLABILITY IN THE PRESENCE OF NOISE

In Chapter I, we considered the problem of reaching a given point $\underline{x}_1$ at a fixed time t_1 with a solution $\underline{\psi}_{\underline{u}}$ of the system I(1) passing through a given point $\underline{x}_0$ at the initial time t_0. No noise was present.

In this chapter we want to allow the system to be subject to noise, entering the system as input possibly through a different path than the input $\underline{u}$.

More precisely, we shall consider the following system

$$\dot{x} = A(t)\underline{x} + B(t)\underline{u} + C(t)\,\underline{n}(t) \qquad \text{III(1)}$$

where $\underline{x}$, $A(t)$, $B(t)$, $\underline{u}$ are defined as in Chapter I, and $C(t)$ is an $n \times h$ matrix whose elements are $\mathcal{L}$-integrable with their q' power in the interval $K = [t_0, t_1]$, and $\underline{n}(t)$ is an h-component noise vector, such that its components are $\mathcal{L}$-integrable with their p'^{th} power, $(p' + q' = 1)$.

Letting again the controls $\underline{u}$ belong to a given class $\mathcal{U}$, the problem we shall consider now is the one of determining conditions under which there exists some $\underline{u} \in \mathcal{U}$ such that the corresponding solution of III(1) stating from $\underline{x}_0$ at t_0 will reach a certain given region A of R^n, for <u>any</u>

noise function $\underline{n}$ belonging to a certain class $\mathcal{N}$.

It is intuitive that in order for the problem to be meaningful, the region A cannot be simply a point $\underline{x}_1$, as it was in Chapter I; indeed, if some $\underline{u}' \epsilon \mathcal{U}$ "transfers" $\underline{x}_0$ at time t_0 into $\underline{x}_1$ at t_1 in the presence of a certain noise $\underline{n}$, the same $\underline{u}'$ cannot effect the same trasfer for an appro priately different noise $\underline{n}'$; applying the same reasoning to all $\underline{x}_1$, $\underline{x}_0$, K , $\mathcal{U}$,the system would not be controllable in the present situation no matter how we choose $\underline{x}_0$, $\underline{x}_1$, K , $\mathcal{U}$.

For semplicity, we shall here choose for A the sphere $A_\varepsilon(\underline{x}_1)$ of radius ε around a given point $\underline{x}_1$:

$$A = A_\varepsilon(\underline{x}_1) = \left\{ \underline{a} : \underline{a} \in R^n, \| \underline{a} - \underline{x}_1 \| \leq \varepsilon \right\} \qquad \text{III(2)}$$

Before facing the problem of controllability in the presence of noise which we mentioned at the beginning, we shall consider again system I(1) (that is system III(1) with $\underline{n} = 0$) and solve the problem of controllability relative to $\underline{x}_0$, $A_\varepsilon(\underline{x}_1)$, $\mathcal{U}$, K . This problem was first formulated and solved by H.Antosiewicz in [1] ; we shall essentially follow here his idea giving a slightly different proof in order to explain in detail and emphasize the classical separation theorem for convex compact sets, which will play an essential role for the solution of the prob lem of controllability in the presence of noise.

We shall equip the controls $\underline{u}$ with the general norm

$$\left|\|\underline{u}(\cdot)\|_r\right|_p = \left(\int_{t_0}^{t_1} \|\underline{u}(s)\|_r^p \, ds\right)^{\frac{1}{p}} \qquad 1 \leq r \leq \infty \qquad 1 < p \leq \infty \qquad \text{III(3)}$$

where

$$\|\underline{u}(t)\|_r = \left(\sum_{1}^{m}{}_i |u_i(t)|^r\right)^{\frac{1}{r}}$$

writing sometimes for convenience $\underline{u}$ instead of $\left|\|\underline{u}(\cdot)\|_r\right|_p$, and let $\mathcal{U} = \mathcal{U}^{\varrho}_{r,p}$ be the set of controls $\underline{u}$ such that $\left|\|\underline{u}\|_r\right|_p \leq \varrho$. In this way, the set $\mathcal{U} = \mathcal{U}^{\varrho}_{r,p}$ becomes a subset of the normed space $\mathcal{L}_{r,p}$ and we may use the important result of Chapter I stating that the reachable set at t_1 :

$$R_{\underline{x}_0}(K) = \left\{ \underline{v}_{\underline{u}}(t_1),\ \underline{u} \in \mathcal{U}^{\varrho}_{r,p} \right\}$$

is convex and compact for $r = 2$ and $1 < p \leq \infty$, since in Chapter I we also mentioned that the same is true for the general case $1 \leq r \leq \infty$, which we are considering here.

Let us now formally state the following

<u>P r o b l e m III,1</u>

Find conditions on the system III(1) with $\underline{n} = 0$ and on $\underline{x}_0$, $\underline{x}_1$, ε , ϱ , K such that there exists at least one control $\underline{u} \in \mathcal{U}$ such that the corresponding solution $\underline{v}_{\underline{u}}$, with ($\underline{v}_{\underline{u}}(t_0) = \underline{x}_0$) reaches some point of $A_\varepsilon(\underline{x}_1)$ at t_1.
We shall call a system III(1) with $\underline{n} = 0$ "$\underline{x}_0\ A_\varepsilon(\underline{x}_1)$, $\mathcal{U}^{\varrho}_{r,p}$, K"

controllable if an $\underline{u}$ such as in Problem III(1) exists. It is clear that the system III(1) (always with $\underline{n} = 0$) is "$\underline{x}_0$, $A_\varepsilon(\underline{x}_1)$, $\mathcal{U}^{\varrho}_{r,p}$, K" controllable if and only if the reachable set $R_{\underline{x}_0}(K)$ at t_1 has a non empty intersection with $A_\varepsilon(\underline{x}_1)$.

The ball $A_\varepsilon(\underline{x}_1)$ is obviously a convex and compact set of R^n, as well as $R_{\underline{x}_0}(K)$; as already mentioned, these properties are basic for the solution of Problem III(1). Let us first give the proof of the separation property of convex compact sets in the form of the following

L e m m a III,1

If C and D are two convex and compact sets of R^n, they are disjoint if and only if there exists an hyperplane strictly separating them; that is, if there exists a vector $\underline{x}'' \in R^{n'}$ and a constant β such that

$$\inf_{\underline{d} \in D} < \underline{x}'', \underline{d} > \;\; > \beta > \sup_{\underline{c} \in C} < \underline{x}'', \underline{c} > \qquad \text{III(4)}$$

Let us notice that we already used a particular case of Lemma III(1), for the proof of the sufficiency part of Theorem I,4; namely, we used inequality I(18), which is a special case of III(4). In Chapter I, inequality I(18) was shown to be a consequence of Lemma I,2.

We shall use here the same Lemma I,2 for the

proof of the present Lemma III(1).

Proof of Lemma III,1

Define the obviously convex and compact set $C = -C'$. The set $D+C' = D-C$ is obviously convex; it is compact (since the function $f(d,c') = d+c'$ is a continuous function on the cartesian product $D \times C' = \{(\underline{d},\underline{c}') \; ; \; \underline{d} \in D, \underline{c}' \in C'\}$, which is compact by Tychonoff's theorem, the set $D+C'$ is the continuous image of a compact set); and, since C and D are disjoint, it does not contain the origin; therefore, by Lemma I,2 there exists a vector $\underline{x}'' \in R^{n'}$ and a positive constant γ such that

$$< \underline{x}'' \underline{y} > > \gamma \qquad \forall \; \underline{y} \in D-C$$

or

$$< \underline{x}'', \underline{d} > - < \underline{x}'', c > \; > \gamma > 0 \qquad \forall \underline{d} \in D \; , \forall \underline{c} \in C$$

and therefore for some β

$$\inf_{\underline{d} \in D} < \underline{x}'', d > \; > \beta > \sup_{\underline{c} \in C} < \underline{x}'', \underline{c} >$$

and Lemma III,1 is proved, since it is obvious that if III(4) holds, C and D are disjoint, and viceversa.

We may now prove the following Lemma III, 2

Lemma III,2

If C and D are two convex compact sets of R^n, a necessary and sufficient condition for them to have a non emp ty intersection is that the following inequality holds:

$$\inf_{\underline{d} \in D} < \underline{x}, \underline{d} > \;\leq\; \sup_{\underline{c} \in C} < \underline{x}, \underline{c} > \quad \forall \underline{x} \in R^{n'} \qquad III(5)$$

P r o o f o f L e m m a III,2

Let C and D have a non-empty intersection $I = C \cap D$, and suppose that there exists an $\underline{f}' \in R^{n'}$ such that

$$\inf_{\underline{d} \in D} < \underline{x}, \underline{d} > \;>\; \sup_{\underline{c} \in C} < \underline{x}, \underline{c} >$$

Then, since $I = C \cap D$ is non-empty, we have

$$\inf_{\underline{i} \in I} < \underline{x}'', \underline{i} > \;\geq\; \inf_{\underline{d} \in D} < \underline{x}'', \underline{d} > \;>\; \sup_{\underline{c} \in C} < x'', \underline{c} > \;\geq\; \sup_{\underline{i} \in I} < \underline{x}'', \underline{i} >$$

and we reach the following contraddiction:

$$\inf_{\underline{i} \in I} < \underline{x}'', \underline{i} > \;>\; \sup_{\underline{i} \in I} < \underline{x}'', \underline{i} > .$$

The necessity is therefore proved.

The sufficiency is obvious by Lemma III,1 since if III(5) holds, the vector $\underline{x}''$ of inequality III(4) cannot exist and therefore C and D are disjoint.

Using Lemma III(2), we are now able to prove the following

T h e o r e m III,1

The system III(1) (with $\underline{n} = 0$) is $\underline{x}_0$, $A_\varepsilon(\underline{x}_1)$, $\mathcal{U}^{\varrho}_{r,p}$, K controllable if and only if for every $\underline{x} \in R^n$ the following inequality holds:

$$<\underline{x},\underline{x}_1 - V(t_1)\underline{x}_0> - \varrho\left(\int_{t_0}^{t_1}\| W^*(t_1,s)\,\underline{x}\|_s^q ds\right)^{\frac{1}{q}} - \varepsilon\|\underline{x}\|^2 \leq 0 \quad \forall \underline{x} \in R^{n'}. \qquad \text{III(6)}$$

where $\frac{1}{p} + \frac{1}{q} = 1$; $\frac{1}{r} + \frac{1}{s} = 1$,

and where, as in Chapter I, $V(t_0)$ is the fundamental matrix solution of the equation

$$\dot{V}(t) = A(t)V(t) \quad \text{with} \quad V(t_0) = I$$

$$W(t_1,s) = V(t_1)V^{-1}(s)B(s)$$

<u>Proof of Theorem III,1</u>

Applying Lemma III,2 to the sets $R_{\underline{x}_0}(K)$ and $A_\varepsilon(\underline{x}_1)$, we may now state that the system III(1) (with $\underline{n}=0$) is $\underline{x}_0$, $A_\varepsilon(\underline{x}_1)$, $\mathcal{U}^{\varrho}_{r,p}$, K controllable if and only if

$$\inf_{\underline{r}\in R_{\underline{x}_0}(K)} <\underline{x},\underline{r}> \;\leq\; \sup_{\underline{a}\in A_\varepsilon(\underline{x}_1)} <\underline{x},\underline{a}> \quad \forall \underline{x}\in R^{n'} \qquad \text{III(7)}$$

In order to have III(6) we need to give a meaningful expression to $\underline{r} \in R_{\underline{x}_0}(K)$, and $\underline{a} \in A_\varepsilon(\underline{x}_1)$, which appear in III(7).

Recalling the definition of $R_{\underline{x}_0}(K)$ and therefore that, as in Chapter I,

$$R_{\underline{x}_0}(K) = \left\{\underline{r} : \underline{r} = V(t_1)\underline{x}_0 + \Lambda_k(\underline{u}),\ \underline{u} \in \mathcal{U}^{\varrho}_{r,p}\right\}$$

where

$$\Lambda_k(\underline{u}) = \int_{t_0}^{t_1} W(t_1, s)\, \underline{u}(s)\, ds \qquad \forall \underline{u} \in \mathcal{U}^{\varrho}_{r,p}$$

we have

$$\inf_{\underline{r} \in R_{\underline{x}_0}(k)} < \underline{x}, \underline{r} > = < \underline{x}, V(t_1)\underline{x}_0 > + \inf_{\underline{u} \in \mathcal{U}^{\varrho}_{r,p}} < \underline{x}, \Lambda_k(\underline{u}) > = < \underline{x}, V(t_1)\underline{x}_0 >$$

$$- \sup_{\underline{u} \in \mathcal{U}^{\varrho}_{r,p}} < \underline{x}, -\Lambda_k(\underline{u}) > = < \underline{x}, V(t_1)\underline{x}_0 > - \sup_{\underline{u} \in \mathcal{U}^{\varrho}_{r,p}} < \underline{x}, \Lambda_k(\underline{u}) > \quad \forall \underline{x} \in R^{n'} \qquad \text{III(8)}$$

since $\mathcal{U}^{\varrho}_{r,p}$ is symmetric with respect to the origin.

Defining $W^*(t_1, s)$ as the adjoint of the matrix $W(t_1, s)$ we have, as in Chapter I,

$$\sup_{\|\underline{u}\| \le \varrho} < \underline{x}, \Lambda_k(\underline{u}) > = \sup_{\|\underline{u}\| \le \varrho} \left| < \underline{x}, \int_{t_0}^{t_1} W(t_1, s)\underline{u}(s) ds > \right| = \sup_{\|\underline{u}\| \le \varrho} \left| \int_{t_0}^{t_1} < W^*(t_1, s)\underline{x}, \underline{u}(s) > ds \right| = \varrho \left(\int_{t_0}^{t_1} \| W^*(t_1, s)\underline{x} \|^q_s \, ds \right)^{\frac{1}{q}} \quad \forall \underline{x} \in R^{n'} \qquad \text{III(9)}$$

Considering the second number of the inequality III(7), we obviously have

$$\sup_{\underline{a} \in A_\varepsilon(\underline{x}_1)} < \underline{x}, \underline{a} > = \sup_{\|\underline{y}\| \le \varepsilon} < \underline{x}, \underline{x}_1 + \underline{y} > =$$

$$< \underline{x}, \underline{x}_1 > + \sup_{\|\underline{y}\| \le \varepsilon} < \underline{x}, \underline{y} > = < \underline{x}, \underline{x}_1 > + \varepsilon \| \underline{x} \|^2 \quad \forall \underline{x} \in R^{n'} \qquad \text{III(10)}$$

Putting together III(7),(8),(9),(10) we have

$$< \underline{x}, V(t_1)\underline{x} > - \varrho \left(\int_{t_0}^{t_1} \| W^*(t_1, s)\underline{x} \|^q_s \, ds \right)^{\frac{1}{q}} \le < \underline{x}, \underline{x}_1 > - \varepsilon \| \underline{x} \|^2 \quad \forall \underline{x} \in R^{n'}$$

or, considering $-\underline{x}$ instead of $\underline{x}$,

$$< \underline{x}, \underline{x}_1 - V(t_1)\underline{x}_0 > - \varrho \left(\int_{t_0}^{t_1} \| W^*(t_1, s)\underline{x} \|^q_s \, ds \right)^{\frac{1}{q}} - \varepsilon \| \underline{x} \|^2 \le 0 \qquad \forall \underline{x} \in R^{n'}$$

which is inequality III(6) and Theorem III,1 is proved.

Behind its intrinsec meaning of giving approximate controllability conditions, i.e. conditions of reachability of a sphere of radius ε around a given point $\underline{x}_1$, Theorem III,1 is interesting to us since its proof may be a guideline for solving the main problem of this chapter, the one of controllability in the presence of noise, i.e. with $\underline{n} \neq 0$ in the system III(1).

We face this problem now. Let us define the class $\underline{n} = \mathcal{N}^{\varrho}_{r',p'}$ of noise functions $\underline{n}$

$$\underline{n} = \left\{ \underline{n} : \Big| \| \underline{n}(s) \|_{r'} \Big|_{p'} \leq \varrho' \right\} \qquad \text{III(11)}$$

where $1 \leq r' \leq \infty$ $1 < p' \leq \infty$, in a way similar to the definition III(2) for $\mathcal{U}^{\varrho}_{r,p}$

We formulate now in a formal way the problem of "controllability in the presence of noise":

Problem III,2

Find conditions on the system III(1) and on $\underline{x}_0$, $\underline{x}_1$, ε , ϱ , ϱ' , K for the existance of at least one control $\underline{u} \in \mathcal{U}$ such that all solutions $v^{\underline{n}}_{\underline{u}}$ of III(1), $(v^{\underline{n}}_{\underline{u}}(t_0) = x_0)$, with any $\underline{n} \in \mathcal{N}$ reach a point of $A_\varepsilon(\underline{x}_1)$ at t_1 .

Let us observe that the values at t_1 of the solutions $v^{\underline{n}}_{\underline{u}}$ of III(1) are of the form

$$v^{\underline{n}}_{\underline{u}}(t_1) = V(t_1)\underline{x}_0 + \int_{t_0}^{t_1} W(t_1, s)\underline{u}(s)\,ds + \int_{t_0}^{t_1} N(t_1, s)\,\underline{n}(s)\,ds$$

with $\underline{u} \in \mathcal{U}^{\rho}_{r,p}$ and $\underline{n} \in \mathcal{N}^{\rho'}_{r',p'}$.

where

$$N(t_1, s) = V(t_1)V^{-1}(s)\,C(s)$$

If we now define the "noise reachable set" $B^{\rho'}_{r',p'}$

$$B = B^{\rho'}_{r',p'} = \int_{t_0}^{t_1} N(t_1, s)\underline{n}(s)ds \qquad \underline{n} \in \mathcal{N}^{\rho'}_{r',p'} \qquad \text{III(12)}$$

we have that to each $\underline{u} \in \mathcal{U}^{\rho}_{r,p}$ there corresponds at t_1 the set of values

$$S_{\underline{u}} = \left\{ V(t_1)\underline{x}_0 + \int_{t_0}^{t_1} W(t_1, s)\,\underline{u}(s)ds + B \right\} \qquad \text{III(13)}$$

We may consider Problem III,2 as the one of finding conditions for the existance of an $\underline{u} \in \mathcal{U}^{\rho}_{r,p}$ such that the corresponding $S_{\underline{u}}$ is <u>entirely</u> contained in $A_\varepsilon(\underline{x}_1)$, or equivalently that

$$\underline{v}_{\underline{u}}(t_1) + B \subset A_\varepsilon(\underline{x}_1) \qquad \text{III(14)}$$

In order to use again the reasoning we followed in Theorem III,1 for the solution of Problem III,1, we define the following set $A'_\varepsilon(\underline{x}_1)$:

$$A'_\varepsilon(\underline{x}_1) = \left\{ \underline{x} : \underline{x} + B \subset A_\varepsilon(\underline{x}_1) \right\}$$

Since $v_{\underline{u}}(t_1)$ which appear in condition III(14) is a member of the set $R_{\underline{x}_0}(K)$, that condition is equivalent to the fact that for some $\underline{x} \in R_{x_0}(K)$, $\underline{x} + B \subset A_\varepsilon(\underline{x}_1)$or, by definition of $A'_\varepsilon(x_1)$,that some $\underline{x} \in R_{\underline{x}_0}(K)$ belongs also to $A'_\varepsilon(\underline{x}_1)$.

In this way we have reduced the problem of controllability in the presence of noise to the condition on the sets $R_{\underline{x}_0}(K)$ and $A'_\varepsilon(\underline{x}_1)$ to have a non empty intersection. We state the above results in the following formal

Statement III,1

System III(1) is $\underline{x}_0$, $A_\varepsilon(\underline{x}_1)$, $\mathcal{U}^{\varrho}_{r,p}$, $\mathcal{N}^{\varrho'}_{r',p'}$, K controllable if and only if

$$R_{\underline{x}_0}(K) \cap A'_\varepsilon(\underline{x}_1) \neq \Phi .$$

At this point, we may use again Lemma III,2 to solve Problem III,2, if we are able to prove that the set $A'_\varepsilon(\underline{x}_1)$ is convex and compact, since we know already that $R_{\underline{x}_0}(K)$ is. Actually, it is easy to prove the following

Lemma III,3

$A'_\varepsilon(\underline{x}_1)$ is a convex and compact set.

Proof

Convexity follows immediately from the definition of B, $A'_\varepsilon(\underline{x}_1)$ and the convexity of $A_\varepsilon(\underline{x}_1)$. Indeed, if $\underline{x}$, $\underline{y} \in A'_\varepsilon(\underline{x}_1)$, then

$$\underline{x} + \underline{b} \subset A_\varepsilon(\underline{x}_1) , \quad \underline{y} + \underline{b} \subset A_\varepsilon(\underline{x}_1) \qquad \forall\, \underline{b} \in B ,$$

and

$$\lambda \underline{x} + (1-\lambda)\underline{y} + \underline{b} = \lambda(\underline{x} + \underline{b}) + (1-\lambda)(\underline{y} + \underline{b}) \in A_\varepsilon(\underline{x}_1) \qquad \forall\, \underline{b} \in B,$$

that is

$$\lambda \underline{x} + (1-\lambda)\underline{y} \in A'_\varepsilon(\underline{x}_1)$$

Compactness is equivalent, since we are in a finite dimensional space, to boundness and closure. $A'_\varepsilon(\underline{x}_1)$ is obviously bounded, since by III(15) it is a subset of $A_\varepsilon(\underline{x}_1)$.

Closure follows from the closure of $A_\varepsilon(\underline{x}_1)$ in a similar way as convexity of $A'_\varepsilon(\underline{x}_1)$ follows from convexity of $A_\varepsilon(\underline{x}_1)$.

Indeed, if $\{\underline{x}_n\}$ is a sequence of $A'_\varepsilon(\underline{x}_1)$ converging to $\underline{x}$, then

$$\underline{x}_n + \underline{b} \in A_\varepsilon(\underline{x}_1) \qquad \forall \underline{n},\ \forall \underline{b} \in B$$

and, since $A_\varepsilon(\underline{x}_1)$ is closed, also $\underline{x} + \underline{b} \in A_\varepsilon(\underline{x}_1)$ for every $\underline{b} \in B$, hence $\underline{x} \in A'_\varepsilon(\underline{x}_1)$.

We may now translate $A'_\varepsilon(\underline{x}_1)$ to a neighborhood of the origin and define the following set

$$A^T_\varepsilon(\underline{x}_1) = A'_\varepsilon(\underline{x}_1) - \underline{x}_1 \qquad \text{III(15)}$$

<u>T h e o r e m   III,2</u>

The system III(1) is $\underline{x}_0$, $A_\varepsilon(\underline{x}_1)$, $\mathcal{U}^\varrho_{r,p}$, $\mathcal{N}^{\varrho'}_{r',p'}$, K controllable if and only if

$$<\underline{x}, \underline{x}_1 - V(t_1)\underline{x}_0> - \varrho\left(\int_{t_0}^{t_1} \| W^*(t_1,s)\underline{x}\|^q_s\, ds\right)^{\frac{1}{q}} - \max_{\underline{y}\in\partial A^T_\varepsilon(\underline{x}_1)} <\underline{x},\underline{y}> \leq 0 \quad \forall \underline{x} \in R^{n'} \qquad \text{III(16)}$$

<u>P r o o f</u>

The proof is simply a repetition of the one giv en for Theorem III,1, taking into account Statement III,1 and observing that

$$\sup_{\underline{a} \in A^T_\varepsilon(\underline{x}_1)} <\underline{x},\underline{a}> = \sup_{\underline{y} \in A^T_\varepsilon(\underline{x}_1)} <\underline{x},\underline{x}_1+\underline{y}> = <\underline{x},\underline{x}_1> + \sup_{\underline{y} \in A^T_\varepsilon(\underline{x}_1)} <\underline{x},\underline{y}>.$$

Since in the last member $<\underline{x},\underline{y}>$ is continuous and linear on $A^T_\varepsilon(\underline{x}_1)$, which is compact by Lemma III,3 and definition III(15), $<\underline{x},\underline{y}>$ assumes its supremum and assumes it on its boundary, that is

$$\sup_{\underline{y} \in A^T_\varepsilon(\underline{x}_1)} <\underline{x},\underline{y}> = \max_{\underline{y} \in \partial A^T_\varepsilon(\underline{x}_1)} <\underline{x},\underline{y}>$$

In order to have a more operative meaning from III(16), it is clear that we need a more detailed expression for $\max\limits_{\underline{y} \in \partial A^T_\varepsilon(\underline{x}_1)} <\underline{x},\underline{y}>$.

For this purpose, let us first prove the following

Lemma III,4

The set $\partial A^T_\varepsilon(\underline{x}_1)$ may be given the following expression:

$$\partial A^T(\underline{x}_1) = \left\{ \underline{y} : \sup_{\|\underline{z}\|=1} <\underline{z},\underline{y}> + \rho'\left(\int_{t_0}^{t_1} \|N(t_1,s)\underline{z}\|^{p'}_{r'} ds\right)^{\frac{1}{p'}} = \varepsilon \right\} \qquad \text{III(17)}$$

Proof

From III(2),(14),(15),

$$A^T_\varepsilon(\underline{x}_1) = \{\underline{y} : \|\underline{y}+\underline{b}\| \le \varepsilon \quad \forall \underline{b} \in B\} = \left\{\underline{y} : \left\|\underline{y} + \int_{t_0}^{t_1} N(t_1,s)\underline{n}(s)\,ds\right\| \le \varepsilon\right\}$$

$$\forall \underline{n} \in \mathcal{n}^{\varrho'}_{r',p'}.$$

that is, by definition of norm of a vector and the linearity of the scalar product:

$$A_\varepsilon^T(\underline{x}_1)=\left\{\underline{y}:\ \sup_{\|z\|=1}\left[<\underline{z},\underline{y}>+<\underline{z},\int_{t_0}^{t_1}N(t_1,s)\underline{n}(s)ds\right]\leq\varepsilon\quad\forall\underline{n}\in\mathcal{N}^{\varrho'}_{r',p'}$$

and therefore also

$$A_\varepsilon^T(\underline{x}_1)=\left\{\underline{y}:\ \sup_{\|z\|=1}\left[<\underline{z},\underline{y}>+\sup_{\underline{n}\in\mathcal{N}^{\varrho'}_{r',p'}}<\underline{z},\int_{t_0}^{t_1}N(t_1,s)\,\underline{n}(s)ds>\right]\right\}\leq\varepsilon=$$

$$\left\{\underline{y}:\ \sup_{\|\underline{z}\|=1}<\underline{z},\underline{y}>+\varrho'\left(\int_{t_0}^{t_1}\|N(t_1,s)\underline{z}\|^{q'}_{s'}\right)^{\frac{1}{q'}}\leq\varepsilon\right\}. \qquad \text{III(18)}$$

where

$$\frac{1}{r'}+\frac{1}{s'}=1 \qquad \frac{1}{p'}+\frac{1}{q'}=1 \qquad 1\leq r'<\infty \qquad 1<p'\leq\infty .$$

Let us now prove that any boundary point $\partial A_\varepsilon^T(\underline{x}_1)$ is given by expression III(17); that is, III(18) with equality sign.

Indeed, take $\underline{y}_0\in\partial A_\varepsilon^T(\underline{x}_1)$ and make the assumption that

$$0<\beta=\left\{\sup_{\|z\|=1}<\underline{z},\underline{y}_0>+\varrho'\left(\int_{t_0}^{t_1}\|N(t_1,s)\,\underline{z}\,\|^{p'}_{r'}\,ds\right)^{\frac{1}{p'}}\right\}<\varepsilon .$$

We may define $\alpha=\dfrac{\varepsilon-\beta}{2}>0$ and take the fol lowing open neighborhood of $\underline{y}_0$:

$$N(\underline{y}_0)=\left\{\underline{y}:\|\underline{y}-\underline{y}_0\|<\alpha\right\};$$

for any $\underline{y}\in N(\underline{y}_0)$, we have

$$\sup_{\|\underline{z}\|=1}\left\{<\underline{z},\underline{y}_0>+\varrho'\left(\int_{t_0}^{t_1}\|N(t_1,s)\underline{z}\|^{p'}_{r'}ds\right)^{\frac{1}{p'}}\right\}\leq$$

$$\sup_{\|\underline{z}\|=1}\left\{<\underline{z},\underline{y}_0>+\varrho'\int_{t_0}^{t_1}\|N(t_1,s)\underline{z}\|^{p'}_{r'}\,ds\right\}+\sup_{\|z\|=1}<\underline{z},\underline{y}-\underline{y}_0>\leq$$

$$\leq \beta + \| y - y_0 \| \leq \beta + \frac{\varepsilon - \beta}{2} = \frac{\beta + \varepsilon}{2} < \varepsilon$$

which means that any $\underline{y} \in N(\underline{y}_0)$ belongs to $A_\varepsilon^T(\underline{x}_1)$, and that is clearly absurd, and Lemma III,4 is proved.

Putting together Theorem III,2 and Lemma III,4 we are now able to state the general and main result of this Chapter, in the form of the following

S t a t e m e n t III,2

The system III(1) is $\underline{x}_0$, $A_\varepsilon(\underline{x}_1)$, $\mathcal{U}^{\rho}_{r,p}$, $\mathcal{N}^{\rho'}_{r',p'}$, K controllable in the presence of noise $\underline{n} \in \mathcal{N}^{\rho'}_{r',p'}$ if and only if

$$< \underline{x}, \underline{x}_1 - V(t_1)\underline{x}_0 > - \rho \left(\int_{t_0}^{t_1} \| W^*(t_1, s)\underline{x} \|_s^q \, ds \right)^{\frac{1}{q}} - \max_{\underline{y} \in \partial A_\varepsilon^T(\underline{x}_1)} < \underline{x}, \underline{y} > \leq 0$$

where the boundary $\partial A_\varepsilon^T(\underline{x}_1)$ is given by

$$\partial A_\varepsilon^T(\underline{x}_1) = \left\{ \underline{y} = \sup_{\| z \| = 1} < \underline{z}, \underline{y} > + \rho' \left(\int_{t_0}^{t_1} \| N(t_1, s)\, \underline{z} \|_{r'}^{p'} \, ds \right) \right\} = \varepsilon$$

At this point the problem of "controllability in the presence of noise" formulated as Problem III,2 is formally solved.

Its specification to particular cases, as for example to specific norms both for $\underline{u}$ and for $\underline{n}$ may be an interesting exercize.

Two important problems remain open to research:

finding appropriate methods in order to get efficient algorithms which translate the above general conditions into numerical and easily workable conditions on $\underline{x}_0$, $\underline{x}_1$, ε , ρ , r , p , ρ' , r' , p' , K , for a system to be "controllable in the presence of noise"; and finding generalizations of the problem to systems which are different from the continuous finite dimensional system we considered here.

REFERENCES

[1] H.A. Autosiewicz:"Linear control systems" Arch. Rat. Mech. Anal. 12, p.313-324 (1963).

[2] W.A. Porter:"Modern foundations of system enegineering" Mc Millan 1966.

[3] N. Dunford, J.T. Schwarz:"Linear operators" part. 1. Inter science pubblishers, 1964.

CONTENTS

Printed in the United States
By Bookmasters